Gary G. Bitter · Jerald L. Mikesell

USING THE MATH EXPLORER™ CALCULATOR

A Sourcebook for Teachers

Addison-Wesley Publishing Company

Menlo Park, California · Reading, Massachusetts · New York
Don Mills, Ontario · Wokingham, England · Amsterdam · Bonn
Sydney · Singapore · Tokyo · Madrid · San Juan

About the Authors

Gary Bitter is Professor of Computer Education and Mathematics at Arizona State University. A teacher and a consultant in the field of computers, mathematics, and curriculum development, Dr. Bitter is also president-elect of the International Society for Technology in Education (ISTE). He is the author of *Computer Literacy,* a basal text for Grades 7-8, and a co-author of *Introduction to BASIC* for middle school students, both published by Addison-Wesley.

Jerald Mikesell is currently Assistant Superintendent for Support Services in the Sierra Vista Public Schools in Arizona. Until recently, Dr. Mikesell was Director of Educational Finance and Support Services for the Alaska State Department of Education. He is a former mathematics teacher.

This book is published by the Addison-Wesley Innovative Division.

Design: Lynn Sanchez

ISBN 0-201-23389-4
BCDEFGHIJKL-AL- 93210

Contents

Preface

Using the Math Explorer™ Calculator: a Sourcebook for Teachers was written to provide the student, teacher, or layperson with a means of becoming familiar with the Math Explorer calculator and its role and use in the curriculum. The reader/participant will attain an understanding and facility in using and applying the Math Explorer. The material in the book can be used as a resource for workshops, inservice courses, individual study, teacher centers, and student calculator curriculum ideas, as a reference for teachers, or as a student resource packet. Numerous activities are available for exploring and understanding the calculator as a tool in the curriculum. The activities are identified by mathematical topic and can be incorporated into any mathematics curriculum.

The Calculator. Most activities require the Texas Instruments Math Explorer pocket calculator.

Instructional Strategies. The background material and the suggested activities are provided for individual, small-group, or large-group instruction. Suggestions for activities appropriate for adult education, PTA groups, in-service training, and primary, intermediate, and secondary instruction are provided. Several inservice workshops are outlined; these include transparency masters. The activities can be used for class presentations, interest centers, mathematics curriculum, or individual activities. These activities can be used for daily or weekly curricula or can be extended over the entire academic year at any particular level. The Texas Instruments Math Explorer calculator is used to teach, explore, discover, and practice various skills in the curriculum. One section includes a discussion of some of the ramifications of the calculator on the curriculum, including considerations for mathematics curriculum development.

Activities. The activities (Chapter 14) are considered the core of this book. Each activity includes the following:

Overview: Each activity has a behaviorally stated objective. The teacher is told what the student should be able to do as a result of successfully performing each activity.

Transparencies: Identifies the transparency masters used from Chapter 2.

Keys Introduced or Used: All the necessary calculator functions are listed.

Student Worksheets: Identifies the student worksheets provided.

Teaching Steps: Step-by-step instruction makes individualization or cooperative learning a reality.

Teacher Helps. Projects that challenge and five-minute fillers are included as resources for establishing a learning climate for using the calculator. The projects that challenge include interdisciplinary application ideas.

Texas Instruments Workshop: A workshop, "Learning to Use the Texas Instruments Math Explorer Calculator," is outlined in Chapter 2, including transparency masters for each key. This workshop can be used for training teachers or for students as each appropriate key is introduced.

Higher-Order Thinking Workshops: The higher-order thinking workshops (Chapters 6-9) include an introduction, which is a review of research, problem-solving strategies, higher-order thinking skills, and relevant curriculum direction. This introduction is followed by primary (K-3), intermediate (4-6), and junior high (7-9) higher-order thinking skills workshops. Each workshop includes several higher-order thinking exercises.

Other Features. Reviews of exponents and scientific notation and estimation with emphasis on place value are among the additional features included in the book.

The authors encourage the use of the calculator as a tool in the curriculum. They feel that the calculator's ability to motivate and stimulate interest is unlimited. Many problems and projects not previously possible due to tedious calculations are now feasible and add realism to the curriculum. Problem solving and higher-order thinking can now be an integral part of the learning of mathematics. Remember, the calculator is a tool, not a curriculum!

1

Introduction

Calculators are continuing to sell in large numbers each year and are becoming available to people of all ages at very low cost. It is now apparent that calculators are making significant inroads into the schools. Students are bringing them to their classes to use, and many school districts are providing calculators for students. It is hoped that teachers will take advantage of the opportunity to expand the horizons of the mathematics curriculum by using calculators. Of course, much is still to be learned of the value and the potential uses of calculators in the school, but their potential seems unlimited. The purpose of this book is to provide ideas for utilizing some of the potential of the Texas Instruments Math Explorer calculator as a tool in the curriculum. It is evident that since calculators are being made available and their use is being encouraged from the home instead of federal grants or projects, their impact will be great. It is hoped that teachers will tap the potential of the Math Explorer as a mathematical tool with the help of this book. But it will take energy, creativity, courage, and a willingness to try to accomplish this goal.

As with any major movement, there is a certain amount of resistance to calculator use in the classroom. But once people see the capabilities and possibilities of the Math Explorer, this resistance should fade and the merits of the calculator should make us want to explore all its potential uses. Most future curriculum recommendations include the calculator as an integral tool to be used by all students.

Many people have heard comments that children will not be able to compute or will be totally dependent on the calculator, and yet research evidence [Hembree, 1986] is available that contradicts this opinion. The potential of using the calculator for realistic problem-solving experiences warrants its use. Consumer and survival math problems can now be solved, with the calculator doing the tedious calculations and the student providing the knowledge to set up the problem so it can be solved. Many adults who oppose children's use of the calculator are totally dependent on calculators in their occupations. In fact, more and more professions are becoming dependent upon calculators.

Through application of the calculator in problem solving and higher-order thinking experiences, people will come to realize that the calculator is a tool, not a replacement for the mathematics curriculum.

For the Individual

For the reader who wants to become knowledgeable and aware of the role and application of the Math Explorer calculator, this book serves as a comprehensive resource that will meet the needs of nearly everyone.

For the Classroom Teacher

There are many calculator activities from which to choose. The activities are organized by common mathematical topics. This makes it possible for the teacher to select activities that complement or enhance existing materials, current textbook series, and curricula. This information permits teachers to integrate these activities with any current mathematics textbook series. The learning activities can be used to meet the needs of individuals and small cooperative groups as well as whole classes. Most of the activities can be used to provide valuable experiences for interest centers or small cooperative group projects.

The activities are structured to some degree, but it is possible to vary from the described procedure and set up alternative ways of solving a given problem. The activities are arranged in such a way as to require the student to solve the experiment or activity in an organized, step-by-step way. Answers are provided. The student is encouraged to experiment to reach a meaningful conclusion or result. The teacher should choose activities in such a way as to establish a meaningful pedagogical sequence. Be sure to allow adequate time for the student to complete the activities. Often activities are too hurriedly done, and the purpose of the activity is not achieved. Encourage experimentation; then discuss answers that at first may not seem to make sense. Through teacher guidance and discussion, errors and misconceptions should gradually be eliminated.

Workshops, Inservice Education, Teacher Centers, Conferences, and Courses

This book is designed to provide calculator activities for use in workshops, inservice education, teacher centers, conferences, and classroom instruction. Inservice workshops are outlined with sample activities in the form of a series of transparency masters included in the book. The transparency masters can be used as models to develop a calculator workshop. The Learning to Use the Math Explorer Workshop includes actual orientation about how to use the calculator. The higher-order thinking skills workshops provide sample problem-solving and higher-order thinking applications. The participants can utilize the calculator to do the activities as they appear on a transparency master, or else the activities can be conveniently set up as interest centers with stations for each category, with participants moving from station to station. Time limits should be established for each rotation. Inservice courses can be handled in a similar way, except that the rotation is on a class or period basis.

General Features

In addition to the activities and workshop chapters, there is a chapter that specifically discusses curriculum considerations, relative to a calculator's use in the classroom as well as its impact on some fundamental mathematics topics now taught. Finally, a brief review of career applications, a chapter on exponents and scientific notation, and a chapter on estimation provide a quick review of some basic mathematics fundamentals needed to use a calculator effectively. The interdisciplinary project ideas, 5-minute fillers, and many other features make this book a valuable resource for teachers who wish to use the Math Explorer calculator in their classrooms.

2

Learning To Use The Math Explorer Calculator

The invention of the integrated circuit eliminated the need for many separate transistors and mechanical connections in electrical designs. Beginning with this invention, technical advances in the calculator industry have been phenomenal. In 1972, Texas Instruments' first standard "calculator-on-a-chip" started today's calculator boom. By 1974, the National Council of Teachers of Mathematics (NCTM) adopted the following position on calculators: "Mathematics teachers should recognize the potential contribution of the minicalculator as a valuable instructional aid. The minicalculator should be used in imaginative ways to reinforce learning and to motivate the learner to become proficient in mathematics." Hundreds of school districts are establishing policies regarding the use of calculators.

In less than 20 years the pocket calculator has gone from dream to reality to being recommended for use in the school. Now, with over 90 percent of all American families owning calculators, it is almost as common to own a calculator as to own a television set. Obviously, the calculator is here to stay.

The calculator is made up of an integrated circuit, circuit board, display, key set, keyboard and solar energy source. The integrated circuit combines all the circuitry in a small chip. The size of the chip is now small enough to allow calculators to be built into watches and pencils. But the size of the calculator must still be large enough for us to be able to use our fingers or a stylus on the keys. The display also must be large enough to be read by the human eye. So, the size is dependent on the person's being able to operate and read the calculator.

The Texas Instruments Math Explorer calculator has many of the function keys other calculators have, such as those used to perform addition, subtraction, multiplication, and division; and percent, constant, memory, square root, pi, backspace, fixed decimal point, and power-of-10 keys. However, only the Math Explorer is able to add, subtract, multiply, divide, and reduce fractions. It also can change from decimals to fractions, and from improper fractions to mixed numbers and back. It can also perform integer division.

The following workshop, Learning to Use the Math Explorer Calculator, is a step-by-step orientation to the Texas Instruments Math Explorer and its capabilities.

Learning To Use the Math Explorer Calculator*

Each calculator has its own distinct architecture for the shape, placement, and type of function keys. Although the differences between calculator function keys are not great, it is still a good practice to familarize oneself with the unique characteristics of a specific calculator. This workshop is designed to demonstrate and provide selected practice on using all the function keys of the Texas Instruments Math Explorer calculator.

Workshop Objectives

The workshop participants should accomplish the following objectives during the course of this workshop:

Identify specific calculator keys and their associated functions.

Use each specific key and function in a practice application.

Practice calculator functions using calculator activities that exercise the functions.

Evaluate the workshop effectiveness for instructing the basic functions of the calculator.

(Use Transparency 1, *Workshop Objectives.*)

*This workshop was developed with a grant from Texas Instruments, and prepared by the Microcomputer Based Learning and Research Laboratory, Arizona State University, Tempe, Arizona, 85287-0111, May, 1988.

Workshop Objectives

Identify calculator keys and their associated functions.

Use each specific key and function in a practical application.

Practice calculator functions using calculator activities that exercise the functions.

Evaluate the workshop effectiveness for instructing the basic functions of the calculator.

Presenter's Guide

There are 42 different keys on the Math Explorer calculator. Their functions can be grouped into the following categories:

1. **Display and maintenance keys:** On/AC, clearing, and backspace keys — 3 keys.
2. **Number keys:** Keys that represent the digits 0 to 9 — 10 keys.
3. **Arithmetic operations keys:** Keys for adding, subtracting, multiplying, dividing, and integer division — 5 keys.
4. **Memory keys:** Keys for entering and retrieving stored information — 3 keys.
5. **Special function keys:** keys for changing signs, decimals, fraction operations, square roots, and powers — 21 keys.

(Use Transparency 2, *Function Key Categories.*)

The standard procedure for progressing through the function key familarization is shown in the ***Quick View*** chart. (See p. 10.) The ***Quick View*** chart can be used by the workshop presenter as a handy reference for sequencing the function key transparency presentation.

Workshop Activities — Overview

The workshop procedure is a simple design. A transparency is furnished for each calculator key function. Practical applications for all major functions are furnished as hands-on activities to use after the explanation and demonstration of the key functions (the applications are called *Try these*).

The use of two overheads can facilitate the workshop by making the keyboard transparency available at all times for reference on the screen as you progress through each function key transparency on another screen.

Function Key Categories

Display and Maintenance—3 Keys
On/AC, clearing, and backspace

Individual Numbers—10 Keys
Keys to represent the digits 0 to 9

Arithmetic Operations—5 Keys
Keys for adding, subtracting, multiplying, dividing, and integer division.

Memory—3 Keys
Keys for entering and retrieving stored information

Special Purpose—21 Keys
Keys for changing signs, decimals, fraction operations, square roots, and powers

Presenter's Guide

Activity 1

The workshop should begin with the overview of the Math Explorer function keys and keyboard layout. A brief orientation to the number display window, solar cells, and keyboard should be done before beginning on the specific function key activities.

(Use Transparency 2, *Function Key Categories,* and Transparency 3, *Keyboard.*)

Activity 2

The procedures, keys to press in order, and the display are shown on each function key transparency in a standardized format. There are transparencies for each specific function key, which include practice exercises. For all function key transparencies, perform the following activities (use transparencies 3 through 27):

1. Demonstrate the function with the participants using the information from each function key transparency in the sequence shown in the *Quick View* chart.
2. Have participants perform the additional practice shown on the transparency in the *Try these* section.
3. Repeat Activity 2 for all function key transparencies.

Activity 3

The final activity is identifying and determining the causes of the error code index. This activity requires the participant to recognize the cause of each error code. Next, the participant is required to reconstruct the erroneous procedures that induce the error codes.

(Use transparencies 23 through 30.)

Time

The workshop is designed to require 4 contact hours of instruction.

Quick View

Presenter Overview

Transparency	Title	Category
3	Keyboard	ON/AC and C/CE
4	ON/AC, Backspace	
5	Math Operations	Number Operations
6	Dividing Whole Numbers	
7	Equivalent Units of Time	
8	Order of Operations	
9	Repeat Operations	
10	Constant Counter	
11	Memory Keys	Memory
12	Percent	Signs
13	Percent Increase or Decrease	Percents
14	Changing the Sign of a Number	Reciprocal
15	Reciprocals	
16	Fixing the Decimal Point	Fractions and
17	Simplifying Fractions: You Choose a Common Factor	Decimals
18	Simplifying Fractions: Calculator Chooses a Factor	
19	Converting Improper Fractions	
20	Math Operations: Fractions	
21	Percents to Fractions	
22	Fractions and Decimals	
23	Powers of 10	Powers and
24	Powers of Numbers	Squares
25	Squares and Square Roots	
26	Parentheses	Formulas
27	Pi	
28	Error Code Index	Errors
29	Error Code Worksheet	
30	Miscalculations	

The Math Explorer Calculator

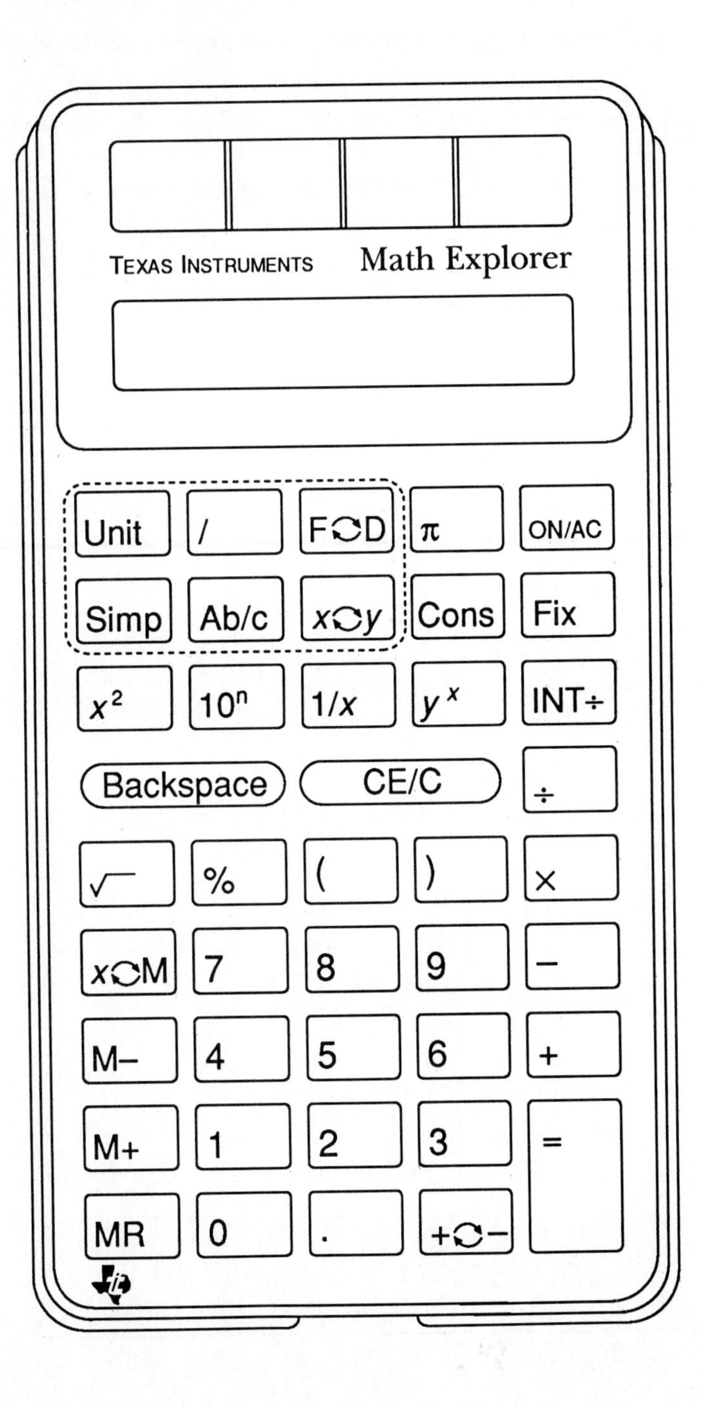

ON/AC, Backspace

Objectives

Identify the procedures for using the [ON/AC] and (Backspace) keys.

Use the [ON/AC] and (Backspace) keys for entering and removing numbers.

Procedure

Use Transparency 3, *Keyboard,* to locate the specific keys if you are using the two-overhead-projector method.

Use Transparency 4, *On/AC, Backspace,* to demonstrate the procedure for turning the calculator on or for clearing a number from the display.

Practice

Have the participants complete the *Try these* problem.

Answer

457

Time

5 minutes

Using the Clearing Keys

ON/AC CE/C Backspace

Procedure	Press	Display
Turn the calculator on; clear the memory.	ON/AC	0
Clear an entry.	CE/C	0
Clear all pending operations.	CE/C	0
Correct an entry error: Enter 1569 and then change it to 1568.	1569 Backspace 8	1569 156 1568

Try these:

Enter 45,790 and remove the last two digits.

Math Operations

Objectives

Identify the procedures in using math operation keys for adding, subtracting, multiplying, and dividing, including the [=] key.

Use the four different math operation keys in conjunction with the [=] key for solving practice problems.

Procedure

Use Transparency 3, *Keyboard,* to locate the specific keys if you are using the two-overhead-projector method.

Use Transparency 5, *Math Operations,* to demonstrate the procedure for performing math operations. Use the example on the transparency to show the use of one type of *math operation* key in conjunction with the [=] key.

Practice

Give the participants several minutes to solve the *Try these* practice problems shown on the transparency.

Answers

a. 148 **c.** 44,765 **e.** 98

b. 3.125 **d.** 336 **f.** 970

Time

5 minutes

Adding, Subtracting, Multiplying, or Dividing Numbers

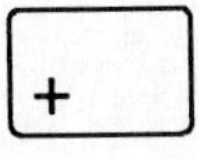
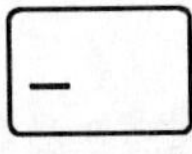
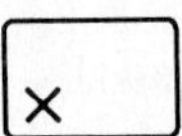
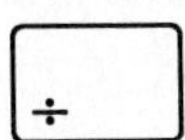

=

Example: 2 + 54 = ?

Press	**Display**
2 [+] 54 [=]	56

Try these:

a. 123 + 25 **c.** 35 x 1279 **e.** 123 - 25

b. 25 ÷ 8 **d.** 16 x 21 **f.** 1204 - 234

Dividing Whole Numbers

Objectives

Identify the procedure for using the function key to divide whole numbers.

Solve practice problems using the whole-number-division function key, [INT÷].

Procedure

Use Transparency 3, *Keyboard*, to locate the specific keys if you are using the two-overhead-projector method.

Use Transparency 6, *Dividing Whole Numbers*, to demonstrate the use of the [INT÷] key. This function key is unique and is not found on other calculators presently in the field. Devote extra care in describing this function key and allow for extensive questioning.

Practice

Give the participants a full 5 minutes to experiment with the function key. Have them solve all the *Try these* practice problems.

Answers

	Q	R		Q	R		Q	R
a.	1	7	**c.**	23	54	**e.**	55	1
b.	0	3	**d.**	3	53	**f.**	2	1

Time

10 minutes

Dividing Whole Numbers

[INT÷]

Example: 173 ÷ 15 = ?

Press	Display
173 [INT÷]	I 173
15 [=]	Q 11 R 8 (quotient of 11, remainder of 8)

Try these:

a. 15 ÷ 8
b. 3 ÷ 5
c. 5436 ÷ 234
d. 242 ÷ 63
e. 111 ÷ 2
f. 7 ÷ 3

Equivalent Units of Time

Objectives

Identify the procedures for calculating equivalent units of time using the [INT÷] key.

Solve practice problems involving hours, minutes, and seconds.

Procedure

Use Transparency 3, *Keyboard,* to locate the specific keys if you are using the two-overhead-projector method.

Use Transparency 7, *Equivalent Units of Time,* to demonstrate how to convert quotients and remainders into hours and minutes, and minutes and seconds.

Practice

Give the participants several minutes to solve the *Try these* practice problems.

Answers

a. 1 minute 7 seconds

b. 3 minutes 54 seconds

c. 0 hours 1 minute 32 seconds

d. 0 hours 3 minutes 51 seconds

Time

5 minutes

Calculating Equivalent Units of Time

[INT÷]

Example: 450 seconds = ? minutes ? seconds

Press	**Display**
450 [INT÷]	I 450
60 [=]	Q 7 R 30 (7 minutes, 30 seconds)

Try these:

a. 67 seconds = __ minutes __ seconds
b. 234 seconds = __ minutes __ seconds
c. 92 seconds = __ hours __ minutes __ seconds
d. 231 seconds = __ hours __ minutes __ seconds

Order of Operations

Objectives

Identify the proper order in which to apply the math operations keys.

Solve practice problems using all four math operations keys following the correct order of operations.

Procedure

Use Transparency 3, *Keyboard,* to locate the specific keys if you are using the two-overhead-projector method.

Use Transparency 8, *Order of Operations,* to demonstrate the sequence for using all four math operations keys.

Practice

Give the participants several minutes to solve the *Try these* practice problems.

Answers

a. 16 **c.** 1300.5

b. 29.5 **d.** 223.88825

Time

5 minutes

Order of Operations

Example: 27 - 5 x 2 = ?

Press	**Display**
27 [−]	− 27
5 [×]	× 5
2 [=]	17

Try these:

a. $4 \times 3 + 6 - 2$

b. $25 + 3^2 \div 2$

c. $1.3 \times 10^3 + .5$

d. $224 - \sqrt{3} \times 2 \div 31$

Repeat Operations

Objectives

Identify the procedure for repeating a math operation function.

Solve practice problems using all math operations keys in a repetitive manner.

Procedure

Use Transparency 3, *Keyboard,* to locate the specific keys if you are using the two-overhead-projector method.

Use Transparency 9, *Repeat Operations,* to demonstrate the repeating of a math operation.

Practice

Give the participants several minutes to solve the *Try these* practice problems.

Answers

a. 512 **b.** 12,800

Time

5 minutes

Repeating an Operation

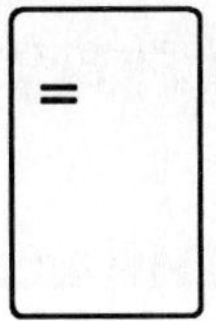

Example: 3 x 2 x 2 x 2 = ?

Press	**Display**
3 [×] 2 [=]	6
[=]	12
[=]	24

Try these:

a. 131,072 ÷ 4 ÷ 4 ÷ 4 ÷ 4

b. 25 x 8 x 8 x 8

Constant Counter

Objectives

Identify the procedure for setting up a constant counter.

Solve practice problems using the constant counter procedure.

Procedure

Use Transparency 3, *Keyboard,* to locate the specific keys if you are using the two-overhead-projector method.

Use Transparency 10, *Constant Counter,* to demonstrate the constant counter process. Use the transparency example and have the participants work with you in setting up the constant counter.

Practice

Give the participants several minutes to solve another problem. Ask them to solve the problem given in *Try these.*

Answer

8 times

Time

5 minutes

Storing a Constant Operation [Cons]

Example: How many times must 4 be subtracted from 12 to get 0?

Procedure	**Press**	**Display**		
Enter the operator.	[–]	-		0
Enter the number you want to add, subtract, multiply, or divide.	4	-		4
Store the constant operation.	[Cons]	CONS		4
Enter the number to be operated upon.	12	CONS		12
Complete the calculation.	[Cons]	CONS	1	8
	[Cons]	CONS	2	4
	[Cons]	CONS	3	0

The counter in the left of the display shows that 4 must be subtracted from 12 three times to get 0.

Try these:

How many times does 5 need to be added to 7 to get 47?

Memory Keys

Objectives

Identify the procedures for entering, adding to, subtracting from, recalling, and clearing the calculator memory.

Solve practice problems using the memory keys.

Procedure

Use Transparency 3, *Keyboard,* to locate the specific keys if you are using the two overhead projector method.

Use Transparency 11, *Memory Keys,* to demonstrate the procedures and sequences for using the memory keys. Use the transparency example and have the participants work along with you in solving problems requiring memory.

Practice

Give the participants several minutes to complete an additional memory problem. Have them complete the *Try these* practice activity.

Answer

105

Time

5 minutes

Using the Memory Keys

[x⇄M] [M+] [M–] [MR]

Procedure	Press	Display
Clear the memory.	[ON/AC]	0
Add the contents in the display to the memory.	25 [M+]	M 25
Subtract the contents in the display from the memory.	16 [M–]	M 16
Recall the memory.	[MR]	M 9
Exchange the contents in the display with the memory.	[ON/AC] 25 [M+] 16 [x⇄M] [x⇄M]	M 0 M 25 M 25 M 16

Try these:

Add 50 to memory. Then add 75 more and subtract 20. Recall the current memory.

Percent

Objectives

Identify the procedures for calculating percentage using the percentage function key.

Solve practice problems using the [%] key.

Procedure

Use Transparency 3, *Keyboard,* to locate the specific keys if you are using the two-overhead-projector method.

Use Transparency 12, *Percent,* to demonstrate how to calculate percentage using the [%] key.

Practice

Give the participants several minutes to solve the *Try these* practice problems.

Answers

a. 44.49925 **c.** 128

b. 1.8612 **d.** 44

Time

5 minutes

Calculating Percentages

[%]

Example: 25% of 42 = ?

Press	**Display**
25 [%] [×]	× 0.25
42 [=]	10.5

Try these:

a. 35.5% of 125.35

b. 12% of 15.51

c. 50% of 256

d. 110% of 40

Percent Increase or Decrease

Objectives

Identify the procedures for calculating the results when a number is increased or decreased by a percentage of itself using the percentage function key.

Solve practice problems requiring the use of the [%] key.

Procedure

Use Transparency 3, *Keyboard,* to locate the specific keys if you are using the two-overhead-projector method.

Use Transparency 13, *Percent Increase or Decrease,* to demonstrate the use of the [%] key for increasing and decreasing with percentages.

Practice

Give the participants several minutes to solve the *Try these* practice problems.

Answers

a. 27.495 **c.** 59.8125

b. 84.0515 **d.** 84

Time

5 minutes

Finding Percent Increase or Decrease

[%]

Example: 170 - 10% = ?

Press	Display
170 [−] 10 [%] [=]	153

Try these:

a. 35.25 - 22%

b. 125.45 - 33%

c. 55 + 8.75%

d. 40 + 110%

Changing the Sign of a Number

Objectives

Identify the procedures for changing the sign of a number.

Solve practice problems that require changing the sign of a number before using it in computation.

Procedure

Use Transparency 3, *Keyboard,* to locate the specific keys if you are using the two-overhead-projector method.

Use Transparency 14, *Changing the Sign of a Number,* to demonstrate procedures for sign changing. Use the transparency example, working along with the participants.

Practice

Give the participants several minutes to solve the *Try these* practice problems.

Answers

a. 3 **c.** -3

b. -84 **d.** -4.6

Time

5 minutes

Changing the Sign of a Number

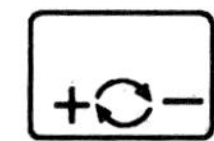

Example: Change 1 to -1 and back to 1.

Press	**Display**
1 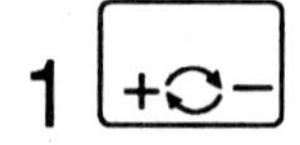	-1
+/−	1

Try these:

a. -1 + 4
b. -7 x 12
c. -6 + 3
d. -23 ÷ 5

Reciprocals

Objectives

Identify the procedure for finding the reciprocal of a number.

Solve practice problems that require determining the reciprocal of a number.

Procedure

Use Transparency 3, *Keyboard,* to locate the specific keys if you are using the two-overhead-projector method.

Use Transparency 15, *Reciprocals,* to demonstrate using the [1/x] key to find the reciprocal of a number. Use at least one example from the transparency with the participants working with you to apply the procedure.

Practice

Give the participants several minutes to solve the *Try these* practice problems.

Answers

a. 0.01 **c.** 0.0588235 **e.** 0.001

b. 0.0192308 **d.** 0.0020202 **f.** 10

Time

5 minutes

Find the Reciprocal

[1/x]

Examples: The reciprocal of 25 = ?
The reciprocal of $\frac{3}{4}$ = ?

Press	Display
25 [1/x]	0.04
3 [/] 4 [1/x]	4/3

Try these:

Find the reciprocal of each number.

a. 100 **c.** 17 **e.** 1000

b. 52 **d.** 495 **f.** .1

Fixing the Decimal Point

Objectives

Identify the procedures for fixing a floating decimal point using the [Fix] key.

Solve practice problems that require establishing fixed decimal points.

Procedure

Use Transparency 3, *Keyboard,* to locate the specific keys if you are using the two-overhead-projector method..

Use Transparency 16, *Fixing the Decimal Point,* to demonstrate fixing a decimal point by using the [Fix] key. Use the transparency example, which shows both how to fix and how to return to a floating decimal point.

Practice

Give participants several minutes to explore fixing decimal points using the *Try these* problems and their own quantities.

Answers

a. 3.142 **b.** 0.083 **c.** 0.142857

Time

5 minutes

Fixing the Decimal Point

Fix

Example: Fix the decimal to four places; then divide 7 by 8.

Procedure	Press	Display
Fix the decimal point. Note: The number of places after the decimal point can be set from 0 to 7.	Fix 4	0.0000
Divide 7 by 8.	7 ÷	÷ 7.0000
	8 =	0.8750
Return to a floating decimal display.	Fix .	0.875

Try these:

a. π, Fix 3 **b.** 1 ÷ 12, Fix 3 **c.** 1 ÷ 7, Fix 6

Simplifying Fractions

Objectives

Identify the procedure for entering fractions into the display of the calculator.

Identify the procedure for simplifying fractions where the user chooses a common factor.

Identify the procedure for simplifying fractions where the calculator chooses a factor.

Solve practice problems requiring the use of the [Simp] and [x⇔y] keys.

Procedure

Use Transparency 3, *Keyboard,* to locate the specific keys if you are using the two-overhead-projector method..

Use Transparencies 17 and 18, *Simplifying Fractions* (you choose, calculator chooses) to demonstrate both methods. Use the example from each transparency to work along with the participants to learn the procedures. (Review or explain the procedure for entering fractions.)

Practice

Give the participants several minutes to solve simplification problems of their own choosing as well as those in the *Try these.* Have them try fractions they feel can and cannot be reduced.

Answers

Transparency 17	**a.** $\frac{3}{8}$	**b.** $\frac{3}{4}$	**c.** $\frac{7}{13}$
Transparency 18	**a.** $\frac{9}{17}$	**b.** $\frac{9}{13}$	**c.** $\frac{7}{9}$

Time

5 minutes

Simplifying Fractions: You Choose a Common Factor

[Simp] [/]

Example: Simplify $\frac{2}{4}$.

Procedure	Press		Display
Enter the fraction.	2 [/] 4		2/4
Prepare to simplify.	[Simp]	SIMP N/D → n/d	2/4
Enter a common factor.	[2]	SIMP	2
Complete the simplification.	[=]		1/2

- If N/D → n/d is no longer in the display, the fraction is simplified.
- If N/D → n/d is still in the display, choose another factor and enter it: [Simp] Enter factor [=].

Try these:

a. $\frac{12}{32}$ **b.** $\frac{27}{36}$ **c.** $\frac{49}{91}$

Simplifying Fractions: Calculator Chooses a Factor

[x⇄y] [/] [Simp]

Example: Simplify $\frac{2}{4}$.

Procedure	**Press**	**Display**
Enter the fraction.	2 [/] 4	2/4
Prepare to simplify.	[Simp]	SIMP N/D → n/d 2/4
Complete the simplification.	[=]	1/2
Optional: Recall the factor.	[x⇄y]	2
Return to the simplified fraction.	[x⇄y]	1/2

- If N/D → n/d is no longer in the display, the fraction is simplified.
- If N/D → n/d is still in the display, press [Simp] [=].

Try these:

a. $\frac{27}{51}$ **b.** $\frac{63}{91}$ **c.** $\frac{91}{117}$

Converting Improper Fractions

Objectives

Identify the procedure for converting improper fractions to a mixed number. (Note that the calculator will display a mixed number by showing first the whole number, then a *u* (for unit), then the fraction.)

Solve practice problems requiring the [Ab/c] and [x⟷y] keys.

Procedure

Use Transparency 3, *Keyboard,* to locate the specific keys if you are using the two-overhead-projector method..

Use Transparency 19, *Converting Improper Fractions,* to demonstrate how to convert improper fractions to mixed numbers. Use the transparency example, working with the participants in group solving of the example.

Practice

Give participants several minutes to solve the *Try these* practice problems.

Answers

a. 3

b. 2 u $\frac{1}{3}$

c. 10 u $\frac{1}{2}$

d. 4

e. 41 u $\frac{1}{4}$

Time

5 minutes

Converting Improper Fractions

[Ab/c] [x⇄y]

Example: Convert $\frac{6}{5}$ to a mixed number.

Procedure	Press	Display
Enter the fraction.	6 [/] 5	6/5
Convert the fraction to a mixed number.	[Ab/c]	1 u 1/5
Optional: Recall the fraction you entered.	[x⇄y]	6/5
Return to the mixed number.	[x⇄y]	1 u 1/5

Try these:

a. $\frac{6}{2}$

b. $\frac{7}{3}$

c. $\frac{21}{2}$

d. $\frac{124}{31}$

e. $\frac{165}{4}$

Math Operations: Fractions

Objectives

Identify procedures for performing math operations with fractions.

Solve practice problems requiring the application of all math operations to fractions.

Procedure

Use Transparency 3, *Keyboard,* to locate the specific keys if you are using the two-overhead-projector method..

Use Transparency 20, *Math Operations: Fractions* to demonstrate at least one example of applying basic operations to fractions.

Practice

Give the participants several minutes to solve the *Try these* practice problems.

Answers

a. $\frac{11}{15}$ **c.** $\frac{5}{8}$

b. $\frac{1}{15}$ **d.** $\frac{3}{4}$

Time

5 minutes

Adding, Subtracting, Multiplying, or Dividing Fractions

[Unit]

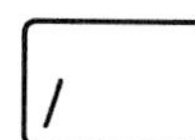

Example: $1\frac{1}{2} - \frac{3}{8} = ?$

Procedure	Press	Display
Enter the whole number.	1 [Unit]	1u
Enter the fraction.	1 [/] 2	1u 1/2
Press the operation.	[–]	-1u 1/2
Complete the calculation.	3 [/] 8 [=]	1u 1/8

Try these:

a. $\frac{1}{3} + \frac{2}{5}$

b. $\frac{2}{5} - \frac{1}{3}$

c. $1\frac{1}{4} + \frac{1}{2}$

d. $\frac{1}{4} \div \frac{1}{3}$

Percents to Fractions

Objectives

Identify the procedures used to change a percentage to a fraction.

Solve practice problems requiring changing a percent to a fraction using the [%], [FOD], and [Simp] keys.

Procedure

Use Transparency 3, *Keyboard,* to locate the specific keys if you are using the two-overhead-projector method.

Use Transparency 21, *Percents to Fractions,* to demonstrate the transformation procedure. Work with the participants to group solve the example shown on the transparency.

Practice

Give the participants several minutes to solve the *Try these* practice problems. Remind participants to simplify the fractions if necessary.

Answers

a. $\frac{1}{8}$ **c.** $\frac{5}{8}$ **e.** $\frac{4}{5}$

b. $\frac{3}{10}$ **d.** $\frac{3}{20}$ **f.** $\frac{41}{500}$

Time

5 minutes

Changing Percents to Fractions

[%]

Example: Write 25% as a fraction.

Press	**Display**
25 [%] [F⇄D]	N/D → n/d 25/100
[Simp] [=] [Simp] [=]	1/4

Try these:

Write each percent as a simplified fraction.

a. $12\frac{1}{2}\%$ **c.** $62\frac{1}{2}\%$ **e.** 80%

b. 30% **d.** 15% **f.** 8.2%

Fractions and Decimals

Objectives

Identify the procedures for changing from a decimal to a fraction.

Solve practical problems using the [F◇D] and [Simp] keys to change fractions to decimals or the reverse.

Procedure

Use Transparency 3, *Keyboard,* to locate the specific keys if you are using the two-overhead-projector method.

Use Transparency 22, *Fractions and Decimals,* to demonstrate changing from fractions to decimals, including simplifying the expression. Use the transparency example to work with the participants in a group setting to solve the example.

Practice

Give the participants several minutes to solve the *Try these* problems.

Answers

a. 3	**b.** 1.4	**c.** 0.75	**d.** 10.875	**e.** 17.25
f. $\frac{1}{2}$	**g.** 2 u $\frac{17}{20}$	**h.** $\frac{1}{20}$	**i.** 6 u $\frac{75}{100}$	**j.** 5 u $\frac{4}{5}$

Time

5 minutes

Changing between Fractions and Decimals

[F⇄D]

Example: Change $\frac{1}{4}$ to a decimal and then back to a fraction.

Procedure	Press	Display
Enter the fraction.	1 [/] 4	1/4
Change the fraction to a decimal.	[F⇄D]	0.25
Change the decimal to a fraction.	[F⇄D]	N/D → n/d 25/100
Simplify.	[Simp] [=]	N/D → n/d 5/20
	[Simp] [=]	1/4

Try these:

Change each fraction to a decimal.

a. $\frac{6}{2}$ **b.** $1\frac{2}{5}$ **c.** $\frac{3}{4}$ **d.** $10\frac{7}{8}$ **e.** $16\frac{5}{4}$

Change each decimal to a simplified fraction.

f. .5 **g.** 2.85 **h.** .05 **i.** 6.750 **j.** 5.8

Powers of 10

Objectives

Identify the procedure used to determine powers of 10.

Solve practice problems involving powers of 10 using the [10^n] key and the [$x \rightleftarrows y$] key.

Procedure

Use Transparency 3, *Keyboard,* to locate the specific keys if you are using the two-overhead-projector method.

Use Transparency 23, *Powers of 10,* to demonstrate how to solve problems involving negative and positive powers of 10. Use the example from the transparency by working with participants in group solving of the examples.

Practice

Give the participants several minutes to solve the *Try these* practice problems.

Answers

a. 700 **b.** 250,000 **c.** 0.074

Time

5 minutes

Using Powers of 10

[10ⁿ] [+⟲−]

Example: $1.3 \times 10^3 = ?$

Press	Display
1 [.] 3 [×]	x 1.3
[10^n] 3 [=]	1300

Example: $1.3 \times 10^{-3} = ?$

Press	Display
1 [.] 3 [×]	x 1.3
[10^n] [+⟲−] 3 [=]	0.0013

Try these:

a. 7×10^2 **b.** 2.5×10^5 **c.** 7.4×10^{-2}

Powers of Numbers

Objectives

Identify the procedures used to calculate the power of a number.

Solve practice problems in calculating powers of a number using the y^x key.

Procedure

Use Transparency 3, *Keyboard,* to locate the specific keys if you are using the two-overhead-projector method.

Use Transparency 24, *Powers of Numbers,* to demonstrate how to calculate the power of a number. Use the transparency example working with the participants as a group.

Practice

Give the participants several minutes to solve the *Try these* practice problems using the y^x key.

Answers

a. 1728 **c.** 1536 **e.** 529

b. 0.2469136 **d.** 0.015625 **f.** 437.89389

Time

5 minutes

Calculating the Power of a Number [y^x]

Example: $5^4 = ?$

Press	**Display**
5 [y^x]	y^x 5
4 [=]	625

Try these:

a. 12^3

b. $22.5 \div 4.5^3$

c. 6×4^4

d. $\left(\frac{1}{4}\right)^3$

e. 23^2

f. $3\left(\frac{3}{8}\right)^5$

Squares and Square Roots

Objectives

Identify procedures used to calculate both the square and the square root of a number.

Solve practice problems calculating squares and square roots using the x^2 and $\sqrt{\ }$ keys.

Procedure

Use Transparency 3, *Keyboard,* to locate the specific keys if you are using the two-overhead-projector method.

Use Transparency 25, *Squares and Square Roots,* to demonstrate calculating squares and roots. Use the transparency examples with the participants in a group solving situation.

Practice

Give the participants several minutes to apply the two keys in solving the *Try these* problems.

Answers

a. 2809 **b.** $\frac{4}{25}$ **c.** 64 **d.** $\frac{9}{11}$

Time

5 minutes

Square a Number x^2
Find the Square Root $\sqrt{\ }$

Examples: $30^2 = ?$

$\left(\frac{1}{3}\right)^2 = ?$

Press	Display
30 [x^2]	900
1 [/] 3 [x^2]	1/9

Examples: $\sqrt{900} = ?$

$\sqrt{\frac{16}{81}} = ?$

Press	Display
900 [$\sqrt{\ }$]	30
16 [/] 81 [$\sqrt{\ }$]	4/9

Try these:

a. 53^2 **b.** $\left(\frac{2}{5}\right)^2$ **c.** $\sqrt{4096}$ **d.** $\sqrt{\frac{81}{121}}$

Parentheses

Objectives

Identify the procedures for using the parentheses keys.

Solve practice problems requiring the use of parentheses in equations.

Procedure

Use Transparency 3, *Keyboard,* to locate the specific keys if you are using the two-overhead-projector method.

Use Transparency 26, *Parentheses,* to demonstrate how the parentheses key is used to simplify expressions. Use the transparency example to work with the participants in a group situation in solving the examples.

Practice

Give the participants several minutes to solve the *Try these* practice problems.

Answers

a. 36 **b.** 90

TIme

5 minutes

Using the Parentheses Keys [( ] [ )]

Example: (27 - 3) x 5 = ?

Procedure	Press	Display
Open the parenthetical expression.	[( ]	() (
Enter the numbers and operation.	27 [−] 3	() 3
Close the parenthetical expression.	[ )]	24
Complete the calculation.	[×] 5 [=]	120

Try these:

a. 3 x (17-5) **b.** (3 + 15) x (14 - 9)

Pi

Objectives

Identify the procedures used to solve circumference problems using the [π] key.

Solve practice problems requiring the use of the [π] key.

Procedure

Use Transparency 3, *Keyboard,* to locate the specific keys if you are using the two-overhead-projector method.

Use Transparency 27, *Pi,* to demonstrate applying the [π] key to circumference problems involving diameters.

Practice

Give the participants about 5 minutes to solve the *Try these* practice problems.The second practice problem will require the participant to use the [Unit] and [/] keys to solve the problem.

Answers

a. 17.592919 inches

b. 10.602875 inches

Time

7 minutes

Pi

π

Example:

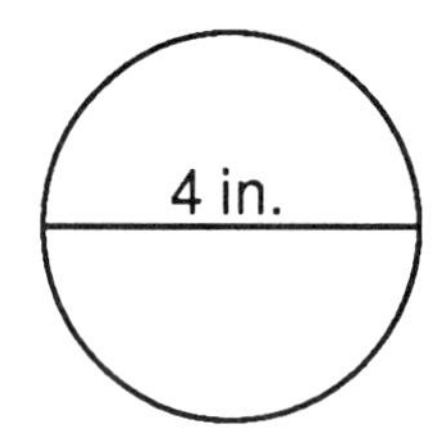

Diameter = 4 inches
Circumference = ?
(π x diameter)

Press	**Display**
	12.566371

Try these:

Find each circumference.

a. Diameter = 5.6 inches

b. Diameter = $3\frac{3}{8}$ inches

Error Code Index

Objectives

Identify the erroneous procedures causing each of the error codes.

Differentiate the erroneous procedures that cause the error codes.

Solve practice problems that generate error codes.

Procedure

Use Transparency 3, *Keyboard,* to locate the specific keys if you are using the two-overhead-projector method.

Use Transparency 28, *Error Code Index,* to identify the types of error codes and the cause of each error. Use Transparency 29, *Error Code Worksheet,* to demonstrate with the participants how to generate the error codes so that they understand both their origins and parameters.

Practice

Give the participants about 10 minutes to generate error codes using their own numbers and keys but using Transparency 29, *Error Code Worksheet,* as a guide for generating the errors.

Answers

None required

Time

25 minutes

Error Code Index

- If you make an entry or try to perform an operation which the calculator cannot accept, it will tell you so with an error code that appears in the display. For example, if you make an arithmetic mistake by trying to divide by zero, you will see the message **Error A** in the display.
- Notice that each error code is different and is self-explanatory. For example, if you make an error while entering fractions, you see **Error F**. If you make an error while using the constant operator, you see **Error C**.
- If you see an error code in the display:
  1. Press CE/C CE/C to clear the calculator.
  2. Re-enter the calculation you were working.

Error Code	Cause
Error A	Arithmetic error. For example, trying to divide a number by zero.
Error C	Constant operator error. Not following the steps for using the constant operator.

Error F	Fraction error. Making an incorrect fraction entry.
Error I	Integer error. Using a non-integer when an integer is required.
Error O	Overflow. Using a number too big for the calculator.
Error P	Parentheses error. Trying to use more than four levels of parentheses or pressing [=] while the "parentheses open" indicator is in the display.
Error S	Sign error. Taking the square root of a negative number.
Error U	Underflow error. Entering a number too small for the calculator.

Error Code Worksheet

	Press	Display
Arithmetic Error	100 [÷] 0 [=] 0 [1/x]	Error A
Constant Operator Error	12 [Cons] 45 [×] [Cons]	Error C
Fraction Error	58 [/] [−] 25 [Simp]	Error F
Integer Error	47 [INT÷] 2.5 [=] 5.8 [INT÷]	Error I
Overflow Error	1,000,000 [×] 100 [=] 1000 [×] [=] [=] [=]	Error O
Parenthesis Error	[(] 10 [−] [=] 25 [+] [)]	Error P

	Press	Display
Sign Error	25 [+⟲−] [√‾] [+⟲−] 25 [+] 36 [+⟲−] [√‾]	Error S
Underflow	0.0000001 [×] 0.0000001 [=]	Error U

Miscalculations

Objectives

Identify the common miscalculations in order of operation functions for both four-function and algebraic operating calculators.

Solve practice problems that cause order of operation miscalculations.

Distinguish between miscalculations caused by four-function calculators and algebraic operating calculators.

Procedure

Use Transparency 3, *Keyboard,* to locate the specific keys if you are using the two-overhead-projector method.

Use Transparency 30, *Miscalculations,* to demonstrate common forms of miscalculations using calculators for math order of operations. Use the transparency examples with participants in a group solution of causes and results of miscalculations. Stress the value of using an algebraic operating calculator as the beginning calculator for students.

Practice

None required

Answers

None required

Time

7 minutes

Miscalculations

Both the four-function and algebraic operating calculator give the same answer for this type of order of operations problem.

2 [×] 6 [+] 5 = ?

Answer: 17

Reverse the order of operations and the calculators give different answers. The answer of the algebraic calculator is correct.

5 [+] 6 [×] 2 [=] ?

Answer: four function: 22

algebraic: 17

Another type of order of operations problem that causes miscalculations is the following. (Again, the algebraic answer is correct.)

2 [×] 4 [+] 6 [×] 7 = ?

Answer: four function: 98

algebraic: 50

This order of operations problem produces results in which the four-function calculator answer is correct and the algebraic calculator is incorrect.

$$\frac{15 \boxed{+} 5}{2}$$

15 [+] 5 [÷] 2 = ?

Answer: four-function: 10

algebraic: 17.5

Workshop Evaluation

Your response to this questionnaire will provide valuable information to the presenter on how effective the workshop is in familarizing you with the calculator. *Please respond to each question carefully. Circle the response that best indicates your opinion concerning the workshop characteristic.*

Presenter	**Poor**			**Excellent**		
1. Knowledge of content	1	2	3	4	5	6
2. Clear presentation	1	2	3	4	5	6
3. Adequately prepared	1	2	3	4	5	6
4. Proper equipment for presentation	1	2	3	4	5	6

Workshop

1. Appropriate content	1	2	3	4	5	6
2. Quality of content material	1	2	3	4	5	6
3. Organization of material	1	2	3	4	5	6

Overall Rating

1. Presenter	1	2	3	4	5	6
2. Workshop	1	2	3	4	5	6

3

The Calculator And The Curriculum

The low cost and easy accessibility of calculators has made their application in the classroom one of the most discussed topics in our schools today. Although opinions vary about how the calculator should be used in the classroom, it is generally recognized that calculators are part of our lives and have a place in our schools.

Whether the calculator becomes as common as the inkwell used to be or if one will be available for every student remains to be seen. Whatever its role, the tremendous motivational and computational possibilities of the calculator make it inevitable that the calculator will have broad application in the mathematics curriculum of the future.

Math drill, real-life applications, problem solving, using higher-order thinking skills and self-checking are commonly recommended uses for calculators in mathematics. But using them in other curriculum areas and for interdisciplinary projects can help make teaching and learning more exciting and rewarding. Because mathematics is part of all disciplines, calculators can be easily and effectively incorporated into projects that combine mathematics, social studies, science, languages, and so on. Of course, in the upper grades it can be used in such areas as home economics, consumer and business mathematics, and shop.

Advantages

The advantages of making calculator use a part of the curriculum are many. Most important, perhaps, calculators are fun to use and give students the possibility of attacking and successfully solving problems that they would not consider attempting otherwise. These may be problems that are beyond the students' present skills or that are overwhelming and bog them down with long, tedious calculations. Often these obstacles have limited the math problems presented to completely fictitious situations that are unrelated to students' lives and their other studies. Any attempt to work with the large or fractional numbers from real-world situations, such as those in a social studies context, was impossible.

With a calculator, long computations can take about the same amount of time and effort as short ones. For example, computations dealing with projections and estimates concerning population or production statistics no longer have to be time-consuming tasks. Students are excited about being able to apply social studies concepts to actual numbers and come up with quick rough estimates from which they can draw conclusions.

Estimation strategies are a vitally important aspect of problem solving that can now be stressed and integrated. Estimation provides students with the opportunity to test rough statistical guesses to find out if ideas are even feasible. Thus, they can use experimentation and discovery processes to develop concepts even before there is complete understanding of algorithmic processes. A child who desires to know the length of a river on a map, for example, could estimate it and then use the map scale and a calculator to compute the actual length as a check of his or her estimate.

Another advantage of using the calculator in the curriculum is that the calculator makes easier the introduction and use of decimals for younger students. Since an understanding of decimals is needed for metric computations, some educators are recommending that decimal computation along with fraction computation be stressed in the middle grades. If this trend continues, which seems likely, use of the calculator can help emphasize powers of 10 earlier and can equip younger children to deal with decimals.

Survival

Through the increased competence and confidence the calculator can bring, students should gain a much better understanding of number relationships. They will be able to apply this "number sense" to everyday situations now and later. The calculator is of particular value in "survival math," an area that includes recognition and knowledge of practical information and skills needed to survive in a changing world. Involving students in calculator activities on consumer and other relevant topics will make them aware that they can solve problems related to themselves and their environment. Calculators make it possible for students — and all of us — to handle situations requiring computation skills without hesitation.

Application and Planning

Effective use of the calculator in the curriculum involves the student in higher-order thinking, inductive and deductive reasoning, sequencing and generalizing skills. Some of these skills will be integrated into content areas naturally by using the calculator as specific, spontaneous opportunities arise, such as predicting the population growth of two countries in a social studies lesson, adding total terms of office for Republican and Democratic presidents from a U.S. history chart, or estimating the number of words in a particular book or essay. Use of the calculator in science experiments is ideal. Students can use calculators to organize and record the collection of data and to make the computations necessary for illustrating the results through graphs and reports.

A variety of calculator uses can be incorporated into broad, interdisciplinary projects. These require a great deal of general planning and careful thought in terms of the ways the calculator can extend the value of the projects. For example, a combination mathematics and physical education project might involve planning a playground obstacle course. Practicality, safety, uniqueness, and available space are major considerations. The calculator can be used to compute the measurements and stability requirements of obstacles. A miniature layout of the course can be created using the calculator for scaling computations. Finally, building-cost estimates can be computed.

The calculator could help in planning a class backpacking trip, combining mathematics, science, geography, and physical education. It could be used to compute distances and elevations and figure out food and water needs.

Mathematics, geography, and social studies textbook lessons can be tied into a study of local industries, such as forestry, mining, or agriculture. The projects can involve the calculator in research on the industries, their impact on the community, and the products produced.

As the use of the calculator in the classroom increases and becomes more sophisticated over the next few years, teachers will develop many new applications as part of their own curricula.

4

Using The Calculator As A Teaching Tool

Parent Concerns

When it is suggested that students be given a calculator early in their educational experience, the initial reaction of many people is often negative. Many fears and arguments against the use are presented: "The students will not learn their basic facts and they will not learn to compute with pencil and paper or mentally" or "They will always rely on punching buttons for the answer." These and similar statements are often given when the subject of using calculators arises.

It is very possible that students will learn to rely too heavily on the calculator if the teacher has not carefully planned for a different outcome. With a carefully planned curriculum and carefully chosen activities, it is possible for the calculator to become a tool in teaching problem-solving strategies and higher-order thinking skills. The well-prepared teacher will have selected examples that show how this can be done. Parents will see that the calculator can be an aid in accomplishing the goals of mastering the understanding of mathematics as well as developing problem-solving strategies and higher-order thinking skills.

Often parents express the fear that their children will become mentally lazy if a calculator is used. They are worried that the discipline of the mind promoted by doing pencil-and-paper mathematics will be missing.

Athough this argument may have some merit, the value of rote memorization of relatively meaningless materials is questionable. If the mind can be stimulated so that the learning of the arithmetic facts is challenging, fun, and meaningful by using the calculator, a great benefit is derived. Often the calculator becomes a means whereby students focus their minds and energy on the creative aspects of problem solving and how to solve the problems rather than concentrating on the arithmetic aspects of the problems.

An example of how the use of calculators can really give meaning to mathematics application and problem solving is illustrated by one teacher showing the applications of geometric formulas. Using pencil-and-paper arithmetic, the students were able to work only two or three problems involving formulas such as $V= \frac{4}{3}\pi r^3$ per day. At the end of the unit of study, the teacher and students were very distressed to find that the students performed very poorly on the test and that they did not know when to apply a formula or which formula to apply. Another time, that same teacher used calculators all through the unit, except at test time. It was found that the students scored much better because they had had an opportunity to apply and work with every formula many times during the unit of study. A great deal of time that had formerly been used up in arithmetic computation was now available for applications. A second benefit was that the students were able to apply the formulas to the world around them because they did not have to keep the numbers involved small and simple so that the computation would be easier and less time-consuming.

Another argument that nearly always comes up is that the student will be helpless if the battery goes dead or the electricity goes off. Since students do need to know their arithmetic facts and how to estimate, they are able to adjust to the inconvenience. Obviously, the student will survive just as we do if the electricity goes off.

The fear that students will become too trusting of the "gadget" is often expressed. It is true that students often take the answer that appears on the display as correct if careful teaching of estimation and number sense is not done. It is important that estimation and number sense become so much a part of working mathematics problems that students can immediately spot answers that are very far from those that are reasonable. Using the calculator makes the teaching of estimation a much more acceptable task because immediate feedback about the viability of the estimate is available. Also, enough time is saved to feel that time spent estimating is justified.

People who study and use computers quickly learn a principle that is just as valid when working with calculators. The saying in the computer field that has real application with calculators is: "garbage in, garbage out." The answer a student gets will be no better than the accuracy with which the numbers and operations are entered into the calculator.

The statement that not everyone will have a calculator or can afford one, giving unfair advantage to some students, should have no application in the classroom. If specific work with a calculator is to be done, provision should be made so that everyone in the class or group has one. If the agreement has been reached that it is permissible to use a calculator to work problems, then every student must have equal access to one before the work is to be turned in.

One of the most valid arguments against the use of the calculator is that with the use of a calculator, the focus of mathematics is the answer only. The technical skills of setting up the problem and the writing of the steps of the solution accurately, neatly, and in the proper order are slighted, and place-value concepts are overlooked because the calculator automatically accounts for place value for the student. These skills are important skills that a student must master because they are important in the study of algebra and higher mathematical topics as well as in many other fields of study. When developing a curriculum that incorporates the calculator as a teaching and learning tool, it is very important that provision be made for the development of these skills.

A positive program of involving parents prior to extensive work with the calculator in the classroom will pay great dividends. It is important to create an opportunity to discuss the concerns expressed here as well as others that the parents list. Illustrations of how these concerns are going to be turned into positive learning experiences will make the parents much more supportive of a mathematics program that uses calculators and will show you to be a teacher who is prepared and ready to do a good job of seeing that their children benefit from the experiences they have in your classroom.

When planning for study sessions with the parents, anticipate the fears and arguments they are going to have. The Introduction to the workshop, Using the Calculator to Develop Higher-Order Thinking Skills, is excellent for reassuring parents. They also need to work through two or three higher-order thinking activities that show how the calculator can strengthen a student's problem-solving abilities. Help them to realize that the calculator can be an asset in learning rather than a liability. The effort made to involve and inform parents will pay big dividends in parental support for you as a teacher and for the total school program.

Educator Concerns

Not since the advent of the "teaching machine" has there been such a controversial teaching aid as is the calculator. Some educators feel it is a gimmick or a crutch and that students are not going to learn mathematics if they are allowed to use the calculator. However, others who have had considerable experience in working with the calculator in the classroom feel it is not a gimmick that will interfere with the learning of mathematics. Instead, they view it as a great motivator, a means of checking work, a verifier of answers, and a needed tool that can make possible learning and applying mathematics through problem solving and higher-order thinking.

Classroom Applications

One of the most widely accepted uses of the calculator in the classroom is checking work previously done by pencil and paper. Several advantages can be listed for this use of the calculator. First, the teacher need not check every problem. A spot check for accuracy and completeness is all that is necessary. Second, immediate feedback and reinforcement are available to the student. The student sees the right answer immediately, instead of having hours or even days pass before getting the feedback needed to know if a correct procedure is being practiced. The calculator not only lets the student know when an answer is wrong but what the right answer is, and at this time one of the very best teaching moments occurs. A third advantage is that the mathematical processes often become more and more clear as students work the steps of the problem and are able to concentrate on what the processes are instead of the rather trivial mathematics facts.

Often the ability to check and rework a problem to get it right before handing it in greatly improves the self-confidence of the student and accuracy of the work. Knowing that the problems they are handing in are correct gives the students a much better self-concept and adds to the pleasure of working problems.

An especially valuable use of the calculator is in debugging problems that have wrong answers. In a multiplication problem, students can learn a great deal about the process of multiplication by finding partial products or about a division problem by finding the partial quotients.

If using the calculator to check pencil-and-paper work is the only viable use of the calculator a teacher can accept, the following procedure is recommended.

1. Introduce calculators to the class after the students have a thorough understanding of the mathematical processes they are to learn. (Allow free time with the calculators so that the novelty of the machine wears off.)
2. Keep calculators under your control, and require students to show a certain number of problems completed, including all steps involved in solving the problems, before they have access to the calculators.
3. When two or more students give different answers to the same problem, let a calculator be the referee.
4. Give calculators to students for drill and practice, but require them to give an oral answer before they can check with calculators. Seeing and reciting correct answers can provide the extra reinforcement that many students need to enable them to master the basic facts.

Motivation Factor

As with any new and different device, there is a certain amount of motivation because of novelty. Time and again it has been demonstrated that even reluctant learners are eager to use calculators. Most teachers who have used calculators in the classroom report that motivation continues as long as students are given interesting things to do with calculators.

Success in using calculators as teaching tools is predicated on a sound educational program. The novelty of the calculator soon wears off, and if the student has learned its value as a mathematical tool, new horizons begin to open up for the student.

For the first time, calculators give the student the power to work really practical problems. For example, a student can find how much the total payment is on a $70,000 home loan with a 10% interest charge compounded annually over 30 years.

Use of calculators for solving problems that occur almost daily can be highly motivating. Being able to determine which size of package is the better buy or shopping with a "ceiling" amount and using a calculator to keep a running total are examples of uses that can be motivational and also make the student a better consumer.

Problem Solving And Higher-Order Thinking Skills

The challenge and reward of finding solutions to problems that were once beyond their grasp can be very motivating to most students. You might try working the following problem with a group of students and see how excited they get when they actually have the means of determining the correct answer. (See "Expansion Team," p. 246, for problems too big for the Math Explorer display.)

Find the solution in the tale about the man who invented the game of chess and asked to be rewarded in wheat. The rate of payment was 1 kernel of wheat for the first square, 2 kernels for the second, 4 for the third, 8 for the next, and so on. How many kernels would he receive for the 64th square? What is the total number of kernels he would receive?

As the calculator becomes a teaching tool, there are some very definite concepts that need to be taught thoroughly. With the calculator, it is possible to have students involved with operations about which they have no understanding. For example, negative answers appear when a larger number is subtracted from a smaller one. Students need to know the significance of the minus sign when it appears.

Students will be curious about the various keys on the calculator and the results that are obtained when they are pressed. Some simple explanation of the various operations, with easy examples from everyday life, will in most cases satisfy the students. A well-planned and challenging use of the calculator using the operations they understand will be reward enough as the students go through the mathematics program. The workshops on higher-order thinking skills illustrate the use of the calculator to develop these skills. Select the workshop appropriate to the abilities of the students or the class. But do not underestimate students' capability to understand concepts when they are not mired down in computation.

Estimation

Using the calculator in the classroom makes it necessary for the teacher to stress the importance of estimating. It is very easy for the student to develop an unreasonable faith in the results shown on calculators. A unit on estimating is essential if students are to become really proficient calculator users. For most people, a calculator has the mystic qualities of a magic box. We do not understand how it works, and perhaps that is not important. But it is important that we know when the answer obtained is reasonable. Chapter 12, "Estimating With Emphasis on Place Value," can be used to develop and review estimation skills.

Decimals

With the extensive use of the calculator in the classroom, it is important for students to become knowledgeable about decimals. They will need to read, write, and understand the significance of decimal notation. Since many of the application problems from everyday experience involve fractions, the ability to change fractions to decimals and vice versa on the calculator is also necessary.

As a student begins to use decimals in computation, a need arises for skill in rounding numbers and recognizing significant digits. Often, students will use all eight digits displayed when only 2- or 3-place accuracy is warranted.

Because of the ease with which students solve problems for which the computation would be beyond their interests or capabilities without a calculator, there is a temptation to allow students to advance too rapidly in solving problems with calculators. Adequate practice and development and some time for internalizing the concepts must be provided, or confusion and disillusionment will be the final outcome of having calculators available. Chapter 13, "Exponents and Scientific Notation," can be used as a review unit.

Place Value

The concept of place value is important in all operations at all levels. Calculators can be used to check the expanded form of a number by listing digits with the appropriate number of zeros after them and then performing the desired operations. Chapter 12, "Estimating With an Emphasis on Place Value," can be used as a review lesson in place value.

Drill And Practice

Calculators can serve the same purpose as flash cards, with quick oral or written responses and immediate reinforcement of the answers.

A Teaching Tool

Until further research is done and a carefully developed curriculum is provided that uses the calculator as the focus in the learning process, it is important to consider curriculum topics that emphasize the problem-solving capabilities of the calculator. Emphasis should be placed on how the calculator can make the learning of concepts more effective and on looking for ways to stress problem solving, higher-order thinking skills, and the application of real-life situations.

Suitable activities for primary-level students involve using the calculator to learn to count and to make up their own problems. At any grade level students can make up problems for themselves within the limits of their abilities and understanding of size, order of numbers, and operations.

As students reach the level of ability that allows the application of formulas to situations, calculators free the students from the very laborious and time-consuming computation that thwarts adequate practice in knowing how and when to apply a particular formula to a problem. Calculators allow students the freedom to concentrate on how and when to apply the correct formula.

Many teachers have used calculators as a motivator for students, allowing them to experiment with them when they have earned some free time. While spontaneous play is needed in the very beginning, students can make profitable use of this free time if provision has been made for some challenging problems, with appropriate recognition for all the students successfully solving the problems.

A calculator can be a very valuable tool, but only to a user who understands the basic ideas, concepts, and meanings behind the instantaneously generated answers it provides. Calculators in the classroom open new and exciting learning possibilities for teachers and students. *Warning:* Once used in your program, calculators are like the proverbial "one peanut." One is not enough.

5

Curriculum Considerations For Use Of The Calculator

Millions of hand-held calculators are sold each year. Projections call for a continually growing market: Calculators will continue to improve, and their capabilities will continue to increase. The popularity of calculators would indicate that the majority of homes have at least one hand-held calculator. With this availability, educators must, more and more, meet the needs of the child who brings a calculator to school expecting to learn how to use it or uses it in class activities. The phenomenon of calculators coming into the school via the home is forcing more teachers to take a second look at the role of calculators and their educational implications. Teachers sometimes try to include them as home games, toys, and show-and-tell in the lower grades, and to some degree teachers are trying to incorporate calculators in a few activities in the intermediate and upper grades. With the availability of calculators to so many students, teachers are beginning to utilize their potential in the curriculum. Calculators are now playing a role in drill and practice exercises, problem solving and estimating, and in consumer applications. During all this time, the calculator's most common use has been as an "answer getter."

Many teachers have had the experience of a student who dislikes mathematics and who wants to do his or her problems at home. Upon some investigation, after assuming that the teacher was responsible for rekindling the child's interest in mathematics, the teacher has found that the student is getting answers to the problems at home with a calculator. As the calculator is used more and more, the mathematics curriculum, being quite hierarchical and definitive, is being carefully reexamined (*Curriculum and Evaluation Standards for School Mathematics*, NCTM, 1989). Teachers, researchers, and curriculum experts are researching the implications calculators have on the curriculum, especially the scope and sequence of mathematics. The following are considerations brought on by the calculator.

Rounding

Calculator computations involving money often have answers of more than two decimal places. Most students are not aware of this, as they see money answers only in dollars and cents (two-place decimals). Therefore, careful development of rounding skills needs to be in the curriculum at earlier grade levels so students can relate to our monetary system. The Math Explorer calculator allows the user to set the number of decimal places given in the answer with the Fix Key.

Fractions to Decimals

The ability to convert fractions to decimals is essential. The formulas $\frac{1}{2}bh$ for the area of a triangle and $\frac{4}{3}$ pi r^3 for the volume of a sphere can be confusing even if a student knows how to use a calculator to multiply. The student needs to know how to represent the fractions $\frac{1}{2}$ and $\frac{4}{3}$ in decimal form. Therefore, the ability to change a fraction to a decimal is an essential skill that the student must have in order to relate to the fraction in its new form. Or, the student needs to be able to consider $\frac{1}{2}bh$ as $\frac{bh}{2}$ or $bh \div 2$; then the student does not need to know that $\frac{1}{2} = 0.5$. The formula $\frac{4}{3}\pi r^3$ can be thought of as $4\pi\frac{r^3}{3}$ if the student chooses, thus illustrating an important mathematics concept. The Math Explorer calculator provides the option of working with fractions and getting fraction solutions. Also, conversion of fractions to decimals and decimals to fractions is a feature.

Decimal Representation of a Fraction

A fraction such as $\frac{1}{6}$ is represented as 0.1666667 on a calculator. Yet involving all seven significant digits in further computation is awkward and unnecessary. The concept of significant digits needs to be developed and used. Our mathematics textbooks show that 0.1666666 can be rounded to 0.16667 or 0.167. However, this rounding affects computational results and can be confusing to students. The Math Explorer calculator allows the user to control the decimal places with the [Fix] key.

Overflow

The product of 1,234,567 × 123,456 is too large for most calculators to display (an overflow results), because calculators usually have displays of eight or fewer digits. Although many calculators indicate that the answer is too large to be handled, techniques need to be developed to solve these "large" problems. Some calculators give the first eight digits; therefore, estimation of 1,234,567 × 123,456 using powers of 10 can enable students to find an approximate answer in situations like these.

Underflow

Answers to problems such as 0.0001 × 0.000056 may also be quite confusing to students because they result in 0 or an underflow error. Many calculators have no means of indicating that the answer is too small to be displayed on the calculator. Students need to be made aware of this problem. It is helpful if the calculators used have a warning for underflow. Also, students need to learn to use scientific notation for solving these kinds of problems. In their study of numbers, students should be taught about "smallness" just as they are taught about "bigness."

Order of Operations

Rules for order of operation need to be emphasized if accurate computation is to occur. Solving the problem $4 + 20 \div 4 + 6$ from left to right gives the answer 12. The correct answer is 15, since you should multiply and divide before you add and subtract (My Dear Aunt Sally rule; see Chapter 2, p. 20). This procedure is overlooked in most calculator applications and often accounts for wrong answers. Although the use of parentheses is emphasized to help solve this type of problem, rules for order of operations need to be included in the mathematics curriculum. Although they are usually included in the upper grades, such as in eighth-grade mathematics and in algebra, they should be included in the lower grades also. The Math Explorer uses the AOS™ (Algebraic Operating System), or algebraic order, so the correct order of operation is automatically adhered to by the calculator. Students need to understand the correct order of operations or they will feel the Math Explorer is wrong; one thing they need to know is situations in which the Math Explorer can be wrong.

Constant Key

The constant key (automatic repetition of a constant addend, multiplier, and so on) allows computation by a number without entering the constant each time (see Chapter 2, p. 24). For example, divide each number 18, 16, 24, and 27 by 9.

[÷] 9 [Cons]	18 [Cons]	2
	16 [Cons]	1.7777778
	24 [Cons]	2.6666667
	27 [Cons]	3

Likewise, multiplication is as follows for $C = \pi D$, where $D = 2, 4, 6$, and 8 and $C = 3.14 \times D$:

[×] 3.14 [Cons]	2 [Cons]	6.28
	4 [Cons]	12.56
	6 [Cons]	18.84
	8 [Cons]	25.12

The Math Explorer calculator's constant key enhances many of these ideas.

Calculator Arithmetic — Internal Counting

Some interesting nonmathematical "arithmetic" problem solving may occur by using the counting feature of the Math Explorer (see Chapter 2, p. 24). For example, for the problem:

37 + 37 + 37

the Math Explorer solution is [+] 37 [Cons] [Cons] [Cons] = 111.

Division by Zero

Some calculators show 5 ÷ 0 = 0. (Many calculators show a flashing zero.) Students need to understand that this solution is incorrect and that any number divided by 0 is undefined. The Math Explorer display reads Error A when such a computation is entered.

The Math Explorer

The Math Explorer calculator has many features designed specifically for education. Therefore, many past curriculum concerns can be alleviated with new technology available. These features are highlighted throughout this book.

6

Workshop: Higher-Order Thinking Skills

Introduction to Workshops Chapters 7, 8, 9

Introductory and Background Materials (Transparency 1)

These materials provide a brief background on calculator development, and on the acceptance of calculators by mathematics educators. The quotes from the National Council of Teachers of Mathematics (NCTM) are potent and may be shared with workshop participants.

Workshop Goals and Agenda (Transparency 2)

This transparency informs participants of what will be accomplished during the workshop.

Summary of Calculator Research (Transparency 3)

This overhead relates calculator research results.

Research Conclusions (Transparency 4)

This transparency relates summary conclusions regarding the efficacy of calculators in education.

Higher-Order Thinking Skills (Transparency 5)

This transparency lists higher-order thinking skills for mathematics as described by the NCTM. The development of these skills is a focus of the NCTM and of this workshop.

Problem-Solving Techniques for Learners (Transparency 6)

This overhead provides a list of important techniques, or "heuristics" (rules of thumb), that learners should employ when solving mathematics problems.

Problem-Solving Techniques for Teachers (Transparency 7)

This overhead lists strategies that teachers can use to promote successful problem solving among students.

Calculators and Higher-Order Thinking (Transparency 8)

This transparency illustrates how calculators allow learners to proceed more efficiently to solutions to higher-order problems. (The term *iteration* used on this transparency and elsewhere refers to a process of successive numerical experimentation and value finding.) The four items listed on this transparency illustrate several ways that a calculator benefits the problem solver.

Introductory and Background Materials

The electronic calculator has become an indispensable tool for our society. Since Texas Instruments first developed calculator-on-a-chip technology in 1972, calculators have become cheap, powerful, and common.

As early as 1974 the National Council of Teachers of Mathematics (NCTM) recognized the potential of calculators in education. In 1987, their support of calculators for mathematics education was put forth even more strongly. The NCTM is equally convinced in its view that mathematics education must shift from emphasizing rote algorithmic skills to higher-order thinking skills.

The following statements are from the 1989 NCTM *Curriculum and Evaluation Standards for School Mathematics:*

> "*The K-4 curriculum should make appropriate and ongoing use of calculators and computers.* Calculators must be accepted at the K-4 level as valuable tools for learning mathematics. Calculators enable children to explore number ideas and patterns, to have valuable concept-development experiences,

to focus on problem-solving processes, and to investigate realistic applications."

"... the teacher's role [will] shift from dispensing information to facilitating learning, from that of director to that of catalyst and coach. The introduction of new topics and most subsumed objectives should, whenever possible, be embedded in problem situations posed in an environment that encourages students to explore, formulate and test conjectures, prove generalizations, and discuss and apply the results of their investigations. Such an instructional setting enables students to approach the learning of mathematics both creatively and independently and thereby strengthen their confidence and skill in doing mathematics."

"The use of technology in instruction should further alter both the teaching and the learning of mathematics.... Calculators and computers with appropriate software transform the mathematics classroom into a laboratory much like the environment in many science classes, where students use technology to investigate, conjecture, and verify their findings. In this setting, the teacher encourages experimentation and provides opportunities for students to summarize ideas and establish connections with previously studied topics."

"Technology, including calculators, computers, and videos, should be used when appropriate. These devices and formats free students from tedious computations and allow them to concentrate on problem solving and other important content. They also give them new means to explore content. As paper-and-pencil computation becomes less important, the skills and understanding required to make proficient use of calculators and computers become more important."

The activities in this workshop demonstrate how calculators provide an indispensible tool in the solution of thought-provoking mathematics problems. The thrust of the workshop is that:

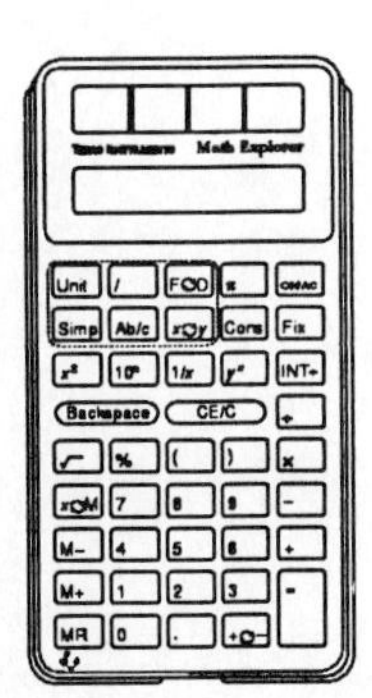

Calculators free the learner to concentrate on the higher-order aspects of a problem instead of the rote computation!

Workshop Goals and Agenda

- *Summarize* recent research findings on the use of calculators in the classroom.
- *Identify* higher-order thinking skills.
- *Recognize* useful problem-solving techniques for learners and teachers.
- *Provide* specific problem-solving activities that are most effectively approached using a hand-held calculator.
- *Discuss* participants' questions and interests regarding calculators and higher-order thinking in mathematics.

Summary of the Research

- The calculator *does not inhibit* learning and knowledge of basic facts.
- The calculator *need not inhibit* teaching pencil-and-paper algorithms.
- Calculators *should be used* to do computations.
- Calculators *should be used* in an iterative manner together with varied examples and nonexamples.
- Calculators *should be used* to study the definitions of operations and concepts.
- Calculators *should be considered* a necessary tool for problem solving.
- Calculators *should be used* for testing.

Hembree, R. and D. J. Dessart, 1986. Effects of hand-held calculators in pre-college mathematics education. *Journal for Research in Mathematics Education.* 17 (2): 83-99.

Research Conclusions

- Calculators *do not prevent* students from learning and understanding math operations, concepts and skills.
- Calculators *do not impair* students' learning and use of mental arithmetic skills, since both students and adults tend to choose the most practical approach for solving math problems.
- The key to using the calculator in classroom teaching is to identify when it provides the most practical approach to the problem.

Shumway, R. 1988. "Calculators and Computers." In *Teaching Mathematics in Grades K-8,* edited by T.R. Post. Allyn & Bacon, Inc., Needham Hts., MA.

Higher-Order Thinking Skills

- Develop Conjectures
- Reason about Phenomena
- Solve Problems
- Build Abstractions
- Validate Assertions

National Council of Teachers of Mathematics, 1989. *Curriculum and Evaluation Standards for School Mathematics*. Reston, Va.: NCTM.

Problem-Solving Techniques for Learners

1. Read

- Find key words
- Understand the problem setting
- Identify what is being asked
- Rephrase the problem

2. Explore

- Draw a picture
- Make a chart
- Record the data
- Look for patterns
- Construct a model

3. Select a Strategy

- Experiment
- Simplify the problem
- Make conjectures
- Form a tentative solution
- Assume a solution

4. Solve

- Implement the strategy
- If no good, go back

5. Critique and Expand

- Verify the solution
- Examine variations

Problem-Solving Techniques for Teachers

- Solve problems in class.
- Encourage students to solve problems.
- Modify problems to involve students and to ensure curricular relevance.
- Encourage unconstrained and creative approaches to problems.
- Let students work in pairs and groups.
- Encourage students to create their own problems.
- Make problems as clear as possible.
- Encourage the use of appropriate tools — mental arithmetic, paper and pencil, and calculator.
- Encourage the use of the problem-solving techniques for learners.

Calculators and Higher-Order Thinking

Many activities illustrate the utility of calculators in solving mathematics problems that require higher-order thinking skills. In several cases, it would be impractical to solve the problems without them. Calculators allow the problem solver to:

Estimate and Validate
Learners can make estimates and immediately confirm the accuracy of their conjectures.

Experiment Iteratively
Learners can proceed toward a solution by using iteration and varying the value of relevant variables.

Concentrate on Strategies
Learners are relieved of the tedium of dull computation and can focus their energy instead on problem solving.

Discover Patterns
By processing numerous values, learners will notice the patterns that emerge as those values vary.

7

Workshop: Using a Calculator to Develop Higher-Order Thinking Skills Grades K-3

Instructions To Presenters

Overview

This workshop is designed to illuminate the ways in which calculators aid problem solvers faced with higher-order math problems. The problem-solving activities included here demonstrate the near necessity of a calculator to:

- Free the problem solver of the burden of manual computation so that he or she may concentrate on higher-order strategies.
- Allow experimentation with varying values to discover patterns and solutions.

The workshop has fully developed problems (Activities 1-5) and supplementary problems (Activities 6-10). The problems in this K-3 workshop are relatively simple. Therefore, presenters should find that the problems take attendees less time to complete. It may be advisable to intersperse details from the front material if the problems are going by too quickly or if participants are bored after several problems. The quotations from the National Council of Teachers of Mathematics (on Transparency 1 in Chapter 6) provide participants with a powerful viewpoint on the importance of calculators and higher-order thinking. All of the quotes refer to the development of a mathematics curriculum. Another way to fill out the presentation would be to have participants create their own problems, or to open the presentation for discussion.

Activity 1:
The Shortest Route

(Transparencies 1 and 2)
This problem requires the summing of whole numbers read from a "map." Learners must understand that the numbers marked between "towns" on the map represent the distance from point to point. The value of using the calculator here is to allow this exercise to become one of visual comprehension and a problem-solving approach rather than being simply an addition problem. Learners may focus on understanding map reading, comparative path evaluation, and the importance of not relying exclusively on visual information. The numbers in conjunction with the visual evidence yield a practical solution. The Extension is a simple extension of the main problem. Simple subtraction (done more efficiently via the calculator) provides the answer.

Mathematical Content

Addition, shortest path, calculator skills

Higher-Order Thinking Skills

Reason about phenomena, solve problems, build abstractions, validate assertions

Potential Problem-Solving Techniques

Understand the problem setting, identify what is being asked, record the data.

Solution to Extension Problem

The extension is solved by subtracting the shortest distance from each of the three longer distances, thus determining the differences:

45 [-] 44 [=] (1 for Route 1)

48 [-] 44 [=] (4 for Route 3)

49 [-] 44 [=] (5 for Route 4)

The Shortest Route

A fire engine needs to get from Point A to Point B . Which route (1, 2, 3, or 4) is the shortest?

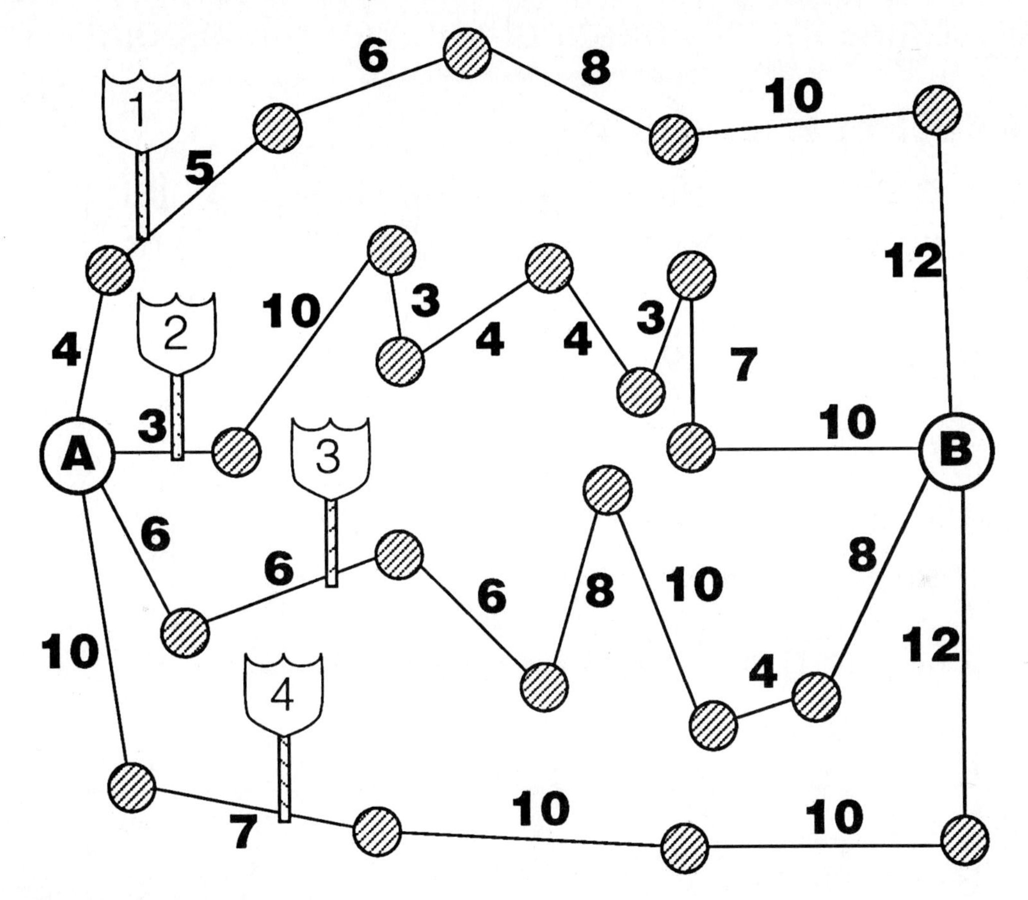

Extension

How much longer is the second shortest route? The third shortest? The longest?

The Shortest Route

Objective

Given a calculator and a "map," learners will determine the shortest path between two points.

Solution Procedure

1. Use the calculator to sum the distance along each route, and write each distance down:

   4 [+] 5 [+] 6 [+] 8 [+] 10 [+]

   12 [=] (45)

   3 [+] 10 [+] 3 [+] 4 [+] 4 [+] 3 [+]

   7 [+] 10 [=] (44)

   6 [+] 6 [+] 6 [+] 8 [+] 10 [+] 4 [+]

   8 [=] (48)

   10 [+] 7 [+] 10 [+] 10 [+] 12 [=] (49)

2. Identify the shortest total distance.

Answer

The shortest distance is along Route 2.

Activity 2: Super Sum

(Transparencies 3 and 4)
This problem involves an understanding of place values. A calculator is helpful for this problem to handle the large number of computations that learners may require to posit and verify a solution. The secret is to realize that to produce the largest sum, the largest digits must be placed in the higher-magnitude decimal places. A strategy is presented in the Solution Procedure on Transparency 4. Although participants should have no difficulty with this problem, it's good to gather candidate solutions from workshop attendees and put them on the board. There are four solutions to the main problem, and participants should be encouraged to discover all of them. The Extension is solved by analogous logic and procedure but with the importance of the place values altered, so that the number being subtracted should be made as small as possible.

Mathematical Content

Place value, addition, maximization, permutation, commutativity, calculator skills

Higher-Order Thinking Skills

Reason about phenomena, develop conjectures, solve problems, validate assertions, build abstractions

Potential Problem-Solving Techniques

Identify what is being asked, make a chart, record the data, look for patterns, make conjectures, form a tentative solution, verify the solution, examine variations

Solution to Extension Problem

The extension problem is solved analogously to the main problem — conjecturing and then using the calculator as an experimental tool to approach the solution iteratively. The largest difference is obtained by subtracting 543 - 12.

Super Sum

Arrange the digits 1 to 5 in the boxes below. Arrange them so that the resulting sum is the largest possible value. You may only use each of the five digits once.

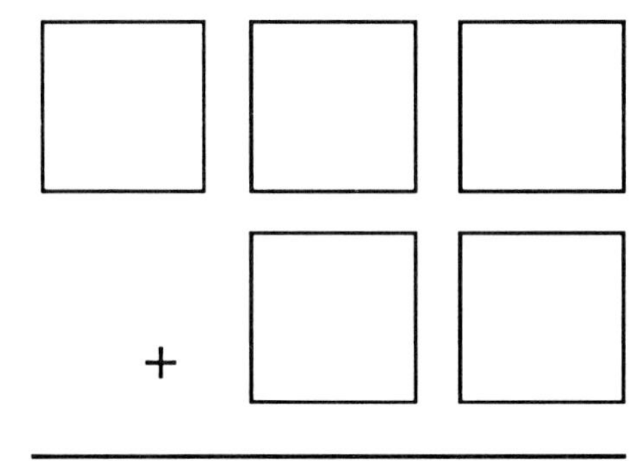

MAXIMUM

Extension

Change the plus sign to a minus sign. Now arrange the five digits to make the resulting difference as large as possible.

Super Sum

Objective

Given a calculator, learners will determine which arrangement of digits produces the greatest sum in a given addition format.

Solution Procedure

1. Place the largest and smallest digits first (5 and 1). After that, the other digits should be placed where they might make the greatest contribution to the sum.
2. This hypothetical "best" arrangement should be tested on the calculator. For instance, a decent guess would be 431 + 52:

   431 [+] 52 [=] (483)
3. Record the numbers and the result. Try other arrangements to see if any is larger (and record the results):

   541 [+] 23 [=] (564)

   542 [+] 31 [=] (573 [the highest value])

Answer

There are four potential correct arrangements: the sum is maximized for 542 + 31, 532 + 41, 531 + 42, and 541 + 32.

Activity 3: Which Plane Flew Furthest?

(Transparencies 5 and 6)
This activity involves a simple application of the formula for computing distance when rate and time are known. Learners do not need to be aware of the formula D=RT to work these problems. They need only to understand the fact that you can figure out how far something traveled by multiplying its speed by a given time traveled. This concept can be presented intuitively to young learners, or they can be directly instructed as to how to compute the distance by simple multiplication of the two numbers presented. In any case, the calculator relieves them of the task of multiplication and instead lets the procedure be their concern. This sort of problem may provide an environment in which youngsters may abstract the formula D=RT. The Extension simply sums the distances traveled by all the aircraft. A calculator again relieves learners of the need to focus on the computational aspect of the problem and instead lets them practice the procedure.

Mathematical Content

Multiplication, D=RT, calculator skills

Higher-Order Thinking Skills

Reason about phenomena, solve problems, build abstractions

Potential Problem-Solving Techniques

Understand the problem setting, identify what is being asked, record the data

Solution to Extension Problem

The extension problem requires learners to sum the distances traveled by all four aircraft:

600 [+] 400 [+] 700 [+] 600 [=] (2300)

Which Plane Flew Furthest?

Various types of aircraft traveled a certain number of hours at a certain speed. The speed and time of travel of each one is listed below. Which plane flew furthest?

A Spitfire flew
3 hours going
200 miles per hour

A biplane flew
4 hours going
100 miles per hour

A jet flew
1 hour going
700 miles per hour

A helicopter flew
4 hours going
150 miles per hour

Extension

What distance was traveled by all the aircraft put together?

Which Plane Flew Furthest?

Objective

Given a calculator, learners will determine distance traveled, given the speed and time of travel.

Solution Procedure

1. Use the calculator to multiply the speed times the hours traveled for each aircraft. Write down each result.

Spitfire:	200 [×] 3 [=]	(600 miles)
biplane	100 [×] 4 [=]	(400 miles)
jet:	700 [×] 1 [=]	(700 miles)
helicopter:	150 [×] 4 [=]	(600 miles)

2. Identify the greatest distance.

Answer

The jet flew furthest: 700 miles.

Activity 4
Addition and Subtraction Targets

(Transparencies 7 and 8)
This is a straightforward problem set. Young learners should approach these problems by successive approximation and calculator verification (though they could be solved by equation solving using inverse operations). Learners should estimate the value of the empty box and then verify their estimates using the calculator. They should progress toward a solution by iteration and successive "zeroing in." This is a problem in which the calculator provides an experimental solution to a problem type before learning equation-solving techniques. The Extension simply allows learners to identify the identity element for addition, 0. (The identity element for an operation maintains the value of the first operand in the operation.)

Mathematical Content

Addition, subtraction, number sentences, calculator skills

Higher-Order Thinking Skills

Develop conjectures, reason about phenomena, solve problems

Potential Problem-Solving Techniques

Identify what is being asked, look for patterns, make conjectures, verify the solution

Solution to Extension Problem

The number 0 can be added to any number and the original number is the result. The problem, if not previously known or solved by insight, can be approached using the calculator to experiment with different values until 0 is recognized as the number in question — the identity element for addition.

Addition and Subtraction Targets

Below are some arithmetic problems. In each problem there is a box where a number should be. What number should be placed in each box to make the computation correct?

a. 34 + ☐ = 47

b. 56 - ☐ = 12

c. 17 + ☐ = 100

d. 73 - ☐ = 14

Extension

What value can you add to 100 so that the sum is still 100? Can you add this number to other numbers besides 100, and get the original number as the result?

Addition and Subtraction Targets

Objective

Given a calculator, learners will fill in a missing operand in an arithmetic computation.

Solution Procedure

1. Learners should proceed via estimation, checking with calculators, and revised estimation to arrive at the solution. For instance, in the second problem, one could proceed as follows:

Try subtracting 50: 56 [−] 50 [=] (6 [too small])

Try subtracting 40: 56 [−] 40 [=] (16 [too big])

Try subtracting 45: 56 [−] 45 [=] (11 [very close])

Try subtracting 44: 56 [−] 44 [=] (12 [correct])

Answers

a. 34 + **13** = 47

b. 56 - **44** = 12

c. 17 + **83** = 100

d. 73 - **59** = 14

Activity 5: No Remainders

(Transparencies 9 and 10)
This is a problem that could be extremely tedious for learners using paper and pencil but is relatively easy to test using the calculator. The calculator allows learners to try as many values as they need to find the greatest divisor of a number. The problem procedure for this problem is pretty straightforward: (1) conjecture, (2) test via computation, and (3) return to step 1, until a tentative solution is found. This common problem type is a prime candidate for calculator solution. The Extension may allow learners to begin to form a generalization about multiples of 10 and the fact that they all end in 0. In fact, any number that ends in 0 is divisible by 10. Again, the calculator permits learners painlessly to test many values in order to ascertain this result.

Mathematical Content

Greatest divisor, division, factors, calculator skills

Higher-Order Thinking Skills

Develop conjectures, reason about phenomena, solve problems, build abstractions, validate assertions

Potential Problem-Solving Techniques

Identify what is being asked, record the data, make conjectures, verify the solution

Solution to Extension Problem

All numbers that end in 0 are evenly divisible by 10. Here, 150, 240, 30, 190, and 110 are all multiples of 10. The common trait is 0 in the ones place on each number, a fact that learners may abstract. A calculator approach is to divide each of the numbers by 10, and observe whether the result is a whole number. For instance, of the first two numbers, 125 is not evenly divisible by 10, but 150 is:

125 [INT÷] 10 [=] (Q 12 R 5 [not a whole number])

150 [INT÷] 10 [=] (15 [Yes, a whole number])

No Remainders

Below are several numbers. Figure out the greatest number that divides evenly (has no remainder) into each of them. You cannot use the number itself.

a. 12 **c.** 35

b. 18 **d.** 45

Extension

Which of the following numbers can be divided evenly by 10? Do they have anything in common?

125 150 240 67 30 190 75 121 110

No Remainders

Objective

Given a calculator, learners will determine the greatest factor of a number (other than the number itself).

Solution Procedure

1. Learners should use estimation and calculator verification. The calculator affords them the luxury of rapid confirmation or rejection of estimates. For instance, to find the greatest factor of 18 in the second problem, learners could reasonably attempt these possibilities:

   18 [INT÷] 2 [=] (9 [all right, but try a larger possible factor])

   18 [INT÷] 3 [=] (6 [all right, but again try a larger possibility])

   18 [INT÷] 4 [=] (Q 4 R 2 [no good, not a factor at all])

Continue this way until 9 is discovered to be the greatest factor.

2. Learners should record the numbers that do divide evenly into each of the numbers given. From this list of factors, they may nominate the largest factor they've found.

3. After several of these problems, learners may begin to recognize certain facts, such as that 2 is a factor of all even numbers, that no number greater than half the number being factored can be a factor of the number, and so on.

Answers

a. The greatest factor of 12 is 6.
b. The greatest factor of 18 is 9.
c. The greatest factor of 35 is 7.
d. The greatest factor of 45 is 15.

Activity 6: A Long Day

(Transparency 11)
This problem is straighforward. Its major attraction will be for youngsters who may find the large numbers interesting. It's an odd way of looking at a day. The only stipulation is that youngsters must understand how many seconds are in a minute, minutes in an hour and so on. They must also understand the power of multiplication in converting between these units of time. The Extension is nearly identical to the main problem. The calculator allows learners to enjoy the large numbers without concern for their generation.

Objective

Given a calculator, learners will determine the number of seconds in a year.

Solution Procedure

Learners must be aware of units of time measurement (60 seconds in a minute, and so on) They may then use the calculator to follow a "conversion path" from seconds to days. This involves consecutive multiplication as follows:

60 [×] 60 [×] 24 [=] (86400)

Answer

There are 86,400 seconds in a day.

Solution to Extension Problem

The extension problem simply adds one more conversion (by multiplication) to the end of the main problem:

60 [×] 60 [×] 24 [×] 365 [=] (31536000)

The result is that there are 31,536,000 seconds in a year.

A Long Day

How many seconds are there in a day?

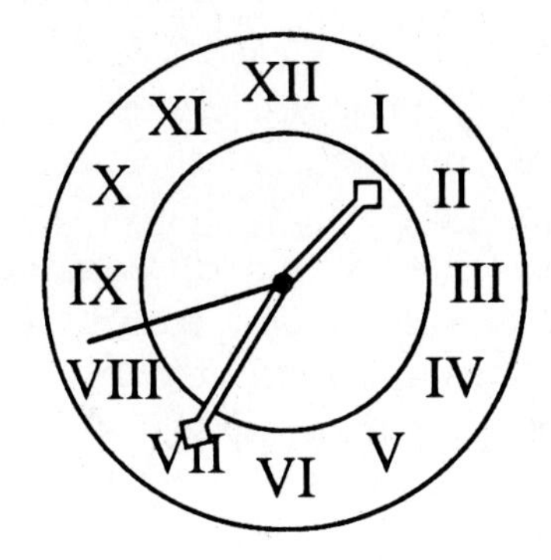

Extension

How many seconds are there in a year?

Activity 7: How Many Stars?

(Transparency 12)
This is a simple problem for the youngest grades. Learners need only count the stars on each of the flags and sum them. A calculator removes the problem from the realm of addition drill and practice and allows learners to examine and enjoy the problem. The Extension is equally simple, requiring only a subtraction of the least amount of stars from the greatest. One advantage of using the calculator on these problems is that learners unskilled at paper and pencil algorithms may participate successfully if they have basic calculator skills.

Objective

Given a calculator, learners will identify (by counting) and sum a series of numbers.

Solution Procedure

Young learners may count the stars and then either write down the count of stars in each flag or use the calculator to keep a running total. In addition to performing the calculation, the calculator can act as a notepad on which the individual tallies are entered. This solution can be done via the following keystrokes:

13 [+] 3 [+] 4 [+] 1 [+] 5 [+] 2 [+] 9 [+] 3 [+] 5 [=] (45)

Answer

There are a total of 45 stars in all the flags.

Solution to Extension Problem

The extension problem requires learners to make note of the number of stars on each flag. At least, they must count the number of stars on the most densely and sparsely starred flags. They must then identify the greatest and least numbers of stars and, finally, subtract the least amount from the greatest:

13 [-] 1 [=] (12)

How Many Stars?

How many stars, altogether, are in the flags below?

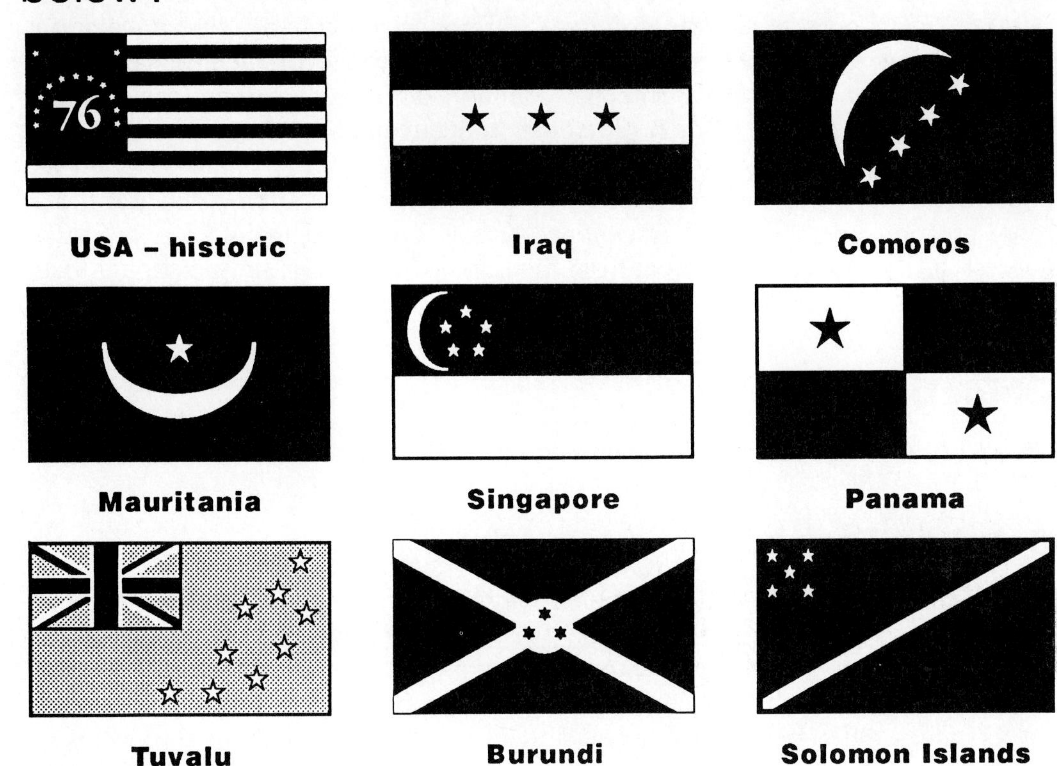

Extension

Identify the flag that has the most stars. Identify the flag that has the fewest stars. How many more stars does the flag with the most stars have than the flag with the fewest stars?

Activity 8: Arithmetic Patterns

(Transparency 13)
This problem involves sequences formed by successive addition of the same number over and over — the common difference. Calculator experimentation permits learners to explore these sequences in advance of their place in the curriculum. After estimating a potential common difference, learners can use the calculator to see if that common difference generates the subsequent term in the progression. If not, they can revise their estimate and verify again as often as needed to reach a tentative solution. The Extension problem is a graphical representation of an arithmetic sequence and is presented mainly for its visual interest. Its solution is no different from the others. In fact, it is simple enough that some learners may be able to compute the missing term mentally.

Objective

Given a calculator, learners will determine missing terms in arithmetic sequences.

Solution Procedure

Some learners may solve these problems by mental arithmetic, but they should verify their solutions on the calculator. Most learners will proceed by estimation and trial and error on the calculator to discern the common difference of each sequence. By establishing the difference between any two numbers in a series, the common difference is ascertained and may then be applied to discover any missing terms. Learners may not know to subtract to find the common difference. In this case, they may proceed by experimental addition to discover what must be added to a specific term to form the next term. For instance, in the third sequence, learners may estimate that the difference must be slightly more than 10. To determine the common difference, try adding different values to form the second of two consecutive terms:

Estimate 10: 45 [+] 10 [=] (55 [too small])

Try 15: 45 [+] 15 [=] (60 [too big])

and so on, until the number 13, the common difference, emerges.

Answers

a. 3	8	13	**18**	23	**28**	33	38
b. 1	11	**21**	31	**41**	51	61	**71**
c. 32	45	58	**71**	84	97	**110**	123

Solution to Extension Problem

The extension problem is a graphic arithmetic progression and is solved in analogous fashion to the main problem. To continue the pattern, there should be 11 fish in the fourth aquarium.

Arithmetic Patterns

Each row of numbers below is an arithmetic sequence. The numbers are generated by adding the same number to the first number over and over again. Fill in the missing numbers so that they match the pattern in that row.

a. 3 8 13 ? 23 ? 33 38
b. ? 11 ? 31 ? 51 61 ?
c. ? 45 58 ? 84 97 ? 123

Extension

How many fish could you expect in Aquarium 4?

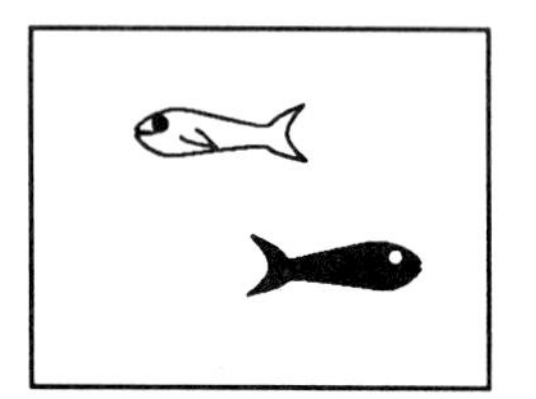
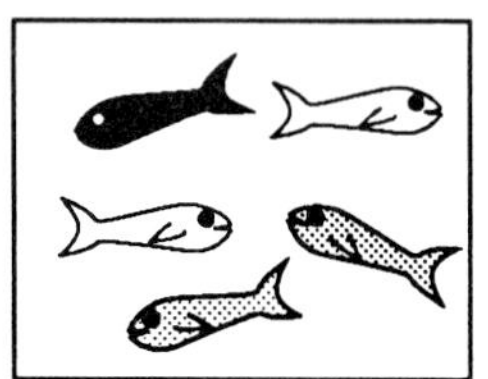
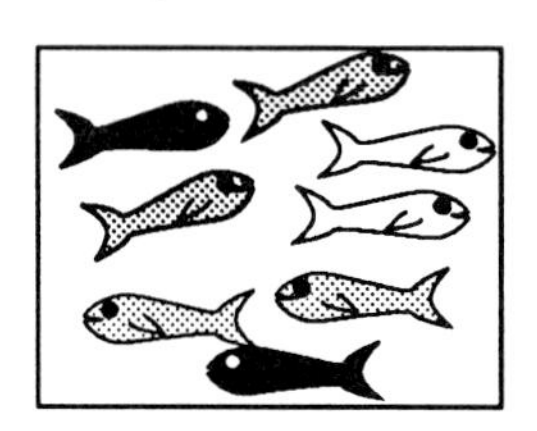

1 **2** **3** **4**

Activity 9: Who Ran the Most Miles?

(Transparency 14)
Here's a problem that requires interpreting a bar graph and then using that information in several sums. The calculator simply provides a convenient computational tool for a problem whose main thrust is graph reading and a problem-solving approach. There is nothing tricky here. The Extension also involves reading the graph and finding a sum. As with most of the problems in this workshop, calculator keystrokes (though simple) are detailed in the solution procedures.

Objective

Given a calculator, learners will interpret a bar graph and perform additions involving the information in the graph.

Solution Procedure

Learners must be capable of interpreting a simple bar graph. From that point, learners must sum information from the graph to arrive at a total for each of the two runners. The calculator provides a tool for computing the running totals. As learners read the daily totals for each runner from the graph, they should record the numbers or simply keep a running total on the calculator. If they write down the information from the graph first, they may subsequently sum the daily figures on the calculator:

Sally: 7 [+] 9 [+] 8 [+] 10 [+] 7 [=] (41)

Joe: 8 [+] 5 [+] 6 [+] 9 [+] 8 [=] (36)

Answer

Sally covered the most miles (41 to Joe's 36).

Solution to Extension Problem

The extension problem requires learners to identify the day on which the most miles were covered by the two runners. Learners may sum the individual mileage of each runner for each day, or they may visually identify several days that could contend for greatest mileage and simply evaluate those days. In any case, the totals for each day are:

Monday: 8 [+] 7 [=] (15)

Tuesday: 5 [+] 9 [=] (14)

Wednesday: 6 [+] 8 [=] (14)

Thursday: 9 [+] 10 [=] (19)

Friday: 8 [+] 7 [=] (15)

The two runners covered the most miles on Thursday.

Who Ran the Most Miles?

Sally and Joe keep track of how far they run each day. The graph below shows how many miles they each ran one week. Who ran farther that week, Sally or Joe?

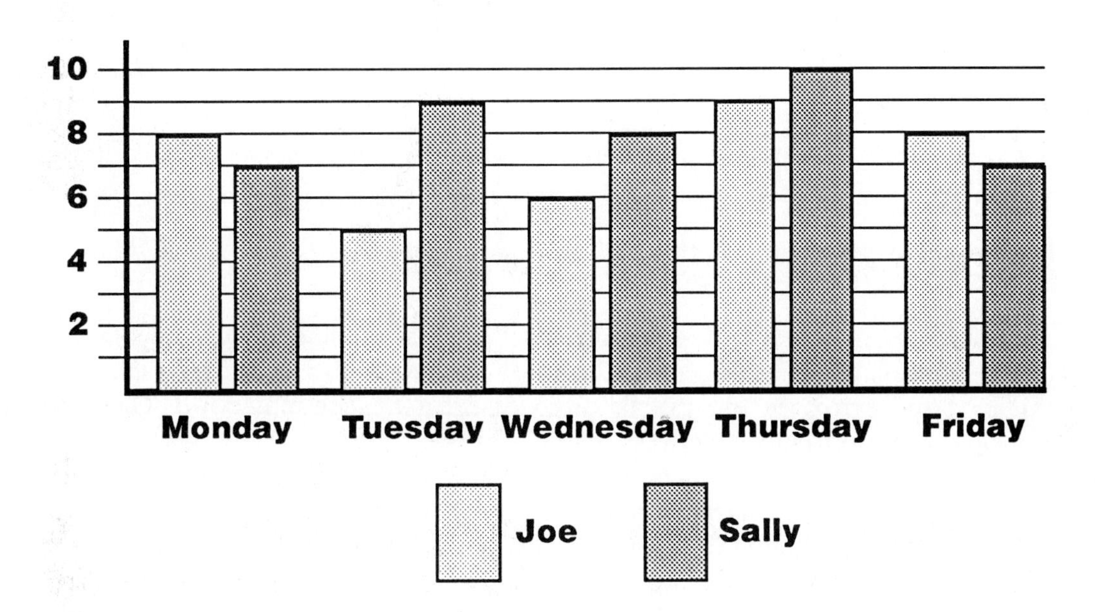

Extension

On which day did they run the most amount of miles altogether?

Activity 10: Multiplication Patterns

(Transparency 15)
This problem is analogous to the Arithmetic Patterns Problem (Transparency 13). After estimating a potential common ratio, learners should use the calculator to see if that common ratio generates the subsequent term in the progression (they must work with two successive terms for which values are given). If not, they can revise their estimate and verify again. The calculator provides a convenient testing facility for learners who are not used to multiplication or working with patterns of numbers. The Extension may be difficult for young learners because they must generate the geometric sequence themselves and then sum it.

Objective

Given a calculator, learners will determine missing terms in simple geometric sequences.

Solution Procedure

Some learners may solve these problems by mental arithmetic, but they should verify their solutions on the calculator. Most learners will proceed by estimation and trial and error via the calculator to discern the common ratio. Once an estimated common ratio is derived, learners can test it out by multiplying one of the terms by the ratio and determining whether the succeeding term matches the next given term (they must choose two consecutive terms whose values are specified). Once a common ratio is found between any two terms, it can, by definition, be used to generate any other terms in the progression. A set of estimates and verifications for the second sequence might be as follows:

Try 10: 5 [×] 10 [=] (50 [much too large])

Try 2: 5 [×] 2 [=] (10 [too small])

Try 4: 5 [×] 4 [=] (20 [the common ratio])

Answers

a. 2	6	18	**54**	162	486
b. 5	20	**80**	320	**1280**	5120
c. 11	11	11	**11**	11	**11**

Solution to Extension Problem

The extension problem involves recognizing a geometric progression whose common ration is 2 and whose first term is 1. Learners must then generate the first five terms of the progression, write them down, and sum them. (The running total could be kept in calculator memory, but this is probably beyond the scope of young learners.) Here are the terms summed:

1 [+] 2 [+] 4 [+] 8 [+] 16 [=] (31)

After 5 days you will have $31.

Multiplication Patterns

Each row of numbers below is a geometric sequence. The numbers are generated by multiplying the first number by the same number over and over again. Fill in the missing numbers so that they match the pattern in that row.

a.	2	6	18	?	162	486
b.	5	20	?	320	?	5120
c.	?	11	11	?	11	?

Extension

Suppose I give you $1 today. Tomorrow I give you twice as much ($2). Every day thereafter, I give you twice as much as the day before. How much (total) will you have after 5 days? (Count the $1 day as the first day.)

8

Workshop: Using a Calculator to Develop Higher-Order Thinking Skills Grades 4-6

Instructions to Presenters

Overview

This workshop is designed to illuminate the ways in which calculators aid problem solvers faced with higher-order math problems. The problem-solving activities included here demonstrate the near necessity of a calculator to:

- Free the problem solver of the burden of manual computation so that he or she may concentrate on higher-order strategies.
- Allow experimentation with varying values to discover patterns and solutions.

The workshop has fully developed problems (Activities 1-5) and supplementary problems (Activities 6-10). The problems in this 4-6 workshop are relatively simple. Therefore, presenters should find that the problems take attendees less time to complete. It may be advisable to intersperse details from the front material if the problems are going by too quickly or if participants are bored after several problems. The quotations from the National Council of Teachers of Mathematics (on Transparency 1 in Chapter 6) provide participants with a powerful viewpoint on the importance of calculators and higher-order thinking. All of the quotes refer to the development of a mathematics curriculum. Another way to fill out the presentation would be to have participants create their own problems, or to open the presentation for discussion.

Activity 1: Leftover Pizzas

(Transparencies 1 and 2)
This problem requires the summing of fractions. A problem that would always have required paper and pencil can be done directly on the Math Explorer calculator. There is nothing tricky about this problem. Learners must look at the diagrams, identify the number of twelfths left at each table, and sum them. If learners were using paper and pencil, it would be best to represent all the fractions as twelfths, sum them, and reduce the result. The calculator allows learners to sum the fractions directly in reduced form, since the calculator will perform fractional sums without preliminary conversion to lowest common denominator. The Extension problem requires learners to reduce the fractional leftovers (if they haven't previously done so). This can be accomplished via inspection or by entering each fraction in twelfths on the calculator and pressing the [Simp] key for automatic reduction (where possible).

Mathematical Content

Representing fractions, addition of fractions, calculator skills

Higher-Order Thinking Skills

Reason about phenomena, solve problems, validate assertions.

Potential Problem-Solving Techniques

Understand the problem setting, identify what is being asked, record the data, form a tentative solution, verify the solution, examine variations

Solution to Extension Problem

The extension may have been solved in the main problem if learners reduced each fractional leftover to simplest form as they were preparing to sum the fractions. Many probably will have worked the main problem using twelfths. The extension requires learners to reduce each fraction from twelfths to simplest form. If learners cannot do this mentally, it may be done easily on the calculator. For example, the fraction of a pizza left on table 1 may be calculated as follows:

3 [/] 12 [Simp] [=] $(\frac{1}{4})$

The fractions are $\frac{1}{4}$, $\frac{1}{2}$, $\frac{1}{12}$, $\frac{1}{6}$, $\frac{5}{12}$, and $\frac{1}{3}$.

Activity 5: What Must the Units Be?

(Transparencies 9 and 10)
This is a relatively easy problem. The trick is to divide the two numbers in each problem in such a way as to produce a recognizable conversion factor. If a division operation is performed in "both directions" with the two numbers in each operation, a recognizable conversion factor is revealed. The applicable conversions are fairly simple, but some learners may need a table of conversion factors. The Extension problem involves a sequence of conversions to go from one unit to another. The details of the most appropriate conversion are presented in the Solution to Extension Problem.

Mathematical Content

Measurement unit conversion, arithmetic operations, calculator skills

Higher-Order Thinking Skills

Reason about phenomena, develop conjectures, solve problems, validate assertions

Potential Problem-Solving Techniques

Identify what is being asked, estimate, experiment, form tentative solutions, verify the solution

Solution to Extension Problem

One mile is farther than 1500 meters (both are race distances for track and field events). Fifteen hundred meters equal approximately 4950 feet (a mile is 5280 feet). The extension problem is a matter of determining a "conversion path" from one of the units to the other. Learners will probably have more success working from meters to miles than from miles to meters. One possible path goes from meters to inches to feet, which can then be compared to the commonly known 5280 feet in a mile.

1500 [×] 39.5 [=] [÷] 12 [=] (4937.5 [feet])

What Must the Units Be?

Below are some "equal measurements." Decide in what units the second measurement must be expressed to make the equality true. (The units may be English or metric.)

a. 144 inches = 12 ________

b. 6 yards = 216 ________

c. 15 000 grams = 15 ________

d. 7 miles = 12,320 ________

Extension

In the Olympics, there are races of 1 mile and 1500 meters. Which is farther?

What Must the Units Be?

Objective

Given a calculator, learners will determine the units involved in given measurement conversions.

Solution Procedure

1. Learners need to know or have access to basic conversion factors (such as inches per yard).

2. Divide the two numbers to find the conversion factor. For instance, in the second problem, dividing 216 by 6 on the calculator yields 36, the number of inches in a yard.

   216 [÷] 6 [=] (36)

3. Proceed in a similar manner to find the conversion factors and the appropriate units for each of the problems on the page. It may be necessary to perform the division in both directions before a conversion factor is recognized. For instance, in the second problem:

   6 [÷] 216 [=] (0.0277778 [not a known conversion factor])

However,

216 [÷] 6 [=] (36 [inches per yard])

Answers

a. 144 inches = 12 feet

b. 6 yards = 216 inches

c. 15 000 grams = 15 kilograms

d. 7 miles = 12,320 yards

Activity 6: Tiling the Floor

(Transparency 11)
This problem requires learners to visualize how tiles would lay over a rectangular floor. It can be approached by envisioning the tiles being laid along two perpendicular edges. Then the number of tiles along each edge can be multiplied to find the total number of tiles required. There are no tricks here, and the calculator serves only to alleviate the need of paper-and-pencil computation. The Extension problem is more interesting because in this case, the half tiles left over from each of the first four rows can be used to fill in the last half-tile requirement at the end of each of the last four rows. Alternatively, the Extension problem can be solved by dividing the area of the floor by the area of a tile, since there are no leftover pieces of tile in this case. Of course, learners must be able to recognize that this approach is viable (some may try this technique to solve the main problem, where it does not work).

Objective

Given a calculator, learners will determine the number of squares of a specified size required to completely cover a specified rectangle.

Solution Procedure

To solve this problem, learners should figure out how many tiles would have to be placed along each of the edges of the floor to cover each edge. Then by multiplying those numbers, the total number of required tiles is discovered. Since the floor is 21 feet long and each tile is 2 feet square, calculate:

21 [÷] 2 [=] (10.5) which means 11 tiles would be needed.

Similarly, it will take 8 tiles to cover the width, since

16 [÷] 2 [=] (8)

Note that $\frac{1}{2}$ tile is left over after each row of 11 tiles is laid. There are 8 rows, so that 8 half tiles are left over.

Answer

Since half tiles that "overhang the other edge" cannot be used, the total number of tiles required is 11 × 8, or 88 tiles.

Solution to Extension Problem

Here, the half tiles that remain can be used again. Therefore, the first 4 rows of tiles laid will provide 4 unused half tiles to finish the last four rows. Therefore, four tiles are saved, requiring a total of 84 tiles. Since no sections of tiles are wasted, this result could have been calculated by dividing the area of the floor by the area of a tile:

21 [×] 16 [=] [÷] 4 [=] (84)

Tiling the Floor

The large rectangle represents a floor. The small square represents a tile. How many tiles will it take to completely cover the floor? (Assume there is no space between the tiles. Also, a tile may be cut to fit, but the leftover piece cannot be used again.)

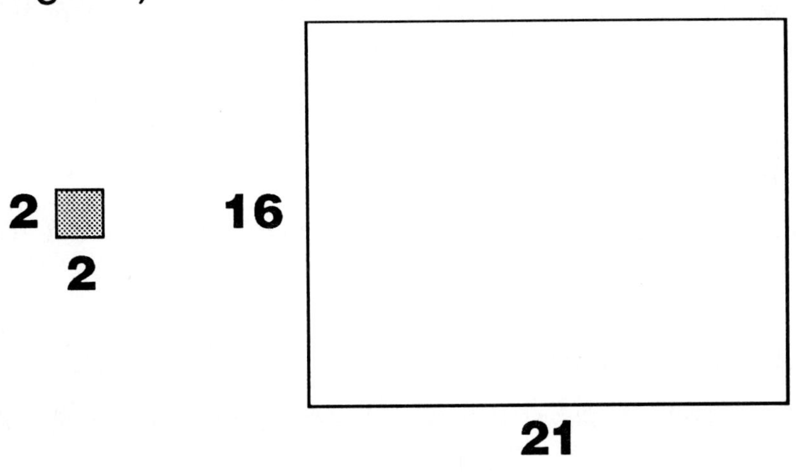

Extension

How many tiles would it take if you could use the pieces of tiles that have been cut?

Activity 7: Addition and Subtraction Targets

(Transparency 12)
This is a straightforward problem set. It can be approached by successive approximation and calculator verification or by equation solving using inverse operations. Learners can either solve the problems directly by "solving for the empty box," as in algebra, or by estimating the value of the empty box and then verifying their estimates using the calculator. They should progress toward a solution by iteration and successive zeroing in. This is a problem in which the calculator provides an experimental solution to a problem type before equation-solving techniques have been learned. The Extension problem has many solutions, but the essential abstraction to be gained is that division by a fraction between 0 and 1 produces a result that is larger than the number being divided.

Objective

Given a calculator, learners will fill in a missing operand in an arithmetic computation.

Solution Procedure

There are two possible approaches here: iterative solution and inverse operation. To solve directly, learners would utilize inverse operations to arrive at the solution. For instance, in the third problem, the inverse operation is division, and the required keystrokes are:

324 [÷] 12 [=] (27)

Learners may also proceed via estimation, calculator checking, and revised estimation to arrive at the solution. For instance, in (a), a learner could proceed as follows after an initial estimate of 200:

234 [+] 200 [=] (434 [too big; try a little smaller])

234 [+] 180 [=] (414 [too small — keep zeroing in])

Answers

a. 234 [+] **191** [=] 425

b. 343 [-] **226** [=] 117

c. 12 [×] **27** [=] 324

d. 666 [÷] **12** [=] 55.5

Solution to Extension Problem

Any number greater than 0 and less than 1 — in other words, any positive proper fraction. Learners can discover and verify this on the calculator using either proper fractions (entered directly as fractions) or with decimal fractions in the same range. For instance, the following represent solutions to the extension problem:

100 [÷] .5 [=] (200)   100 [÷] 1 [/] 2 [=] ($\frac{200}{1}$)

Addition and Subtraction Targets

Below are some arithmetic problems. In each problem there is a box where a number should be. What number should be placed in each box to make the computation correct?

a. 234 + ☐ = 425

b. 343 - ☐ = 117

c. 12 x ☐ = 324

d. 666 ÷ ☐ = 55.5

Extension

What number could you divide 100 by to get an answer that is bigger than 100?

Activity 8: Arithmetic Sequences

(Transparency 13)
This problem is straightforward. Learners should use the [Cons] key of the calculator to regenerate the arithmetic progression problems. After estimating a potential common difference, learners can use the calculator to see if that common difference generates the subsequent values in the progression. If not, they can revise their estimate and verify again. The necessary keystrokes to generate an entire progression are detailed in the Solution Procedure for this problem. The keystrokes for the Extension problem are detailed in the Extension Solution.

Objective

Given a calculator, learners will determine missing terms in arithmetic sequences.

Solution Procedure

Some learners may solve these problems by mental arithmetic, but they should verify their solutions on the calculator. Most learners will proceed by estimation and trial and error on the calculator to discern the common difference of each sequence. The [Cons] key can be helpful in generating the sequence. Refer to the Solution to Extension Problem for the keystrokes required to proceed in this manner.

Answers

a.	3	17	31	**45**	59	73	**87**	101
b.	134	111	**88**	65	**42**	19		
c.	**97**	84	**71**	58	45	32	19	**6**
d.	$0\frac{1}{4}$	$\frac{1}{2}$	$\frac{3}{4}$	1	$1\frac{1}{4}$	$\mathbf{1\frac{1}{2}}$	$1\frac{3}{4}$	2

Solution to Extension Problem

The student will have $198 after the 14th week. The extension problem can be solved iteratively by CONStant addition of 12 to the original $30. The Math Explorer calculator displays the number of iterations completed when the total reaches $198. This is accomplished by pressing [+] 12 [Cons] and then 30 [Cons] [Cons] [Cons]... until the total reaches 198. The number of iterations required to reach that amount (14) appears on the display. An alternate solution is to divide the difference between $198 and $30 ($168) by 12, with a result of 14 weeks required to save the total amount needed.

Arithmetic Sequences

Each row of numbers below is an arithmetic sequence (formed by consecutive addition or subtraction of the same number). Fill in the missing numbers so that they match the pattern in that row.

a. 3 17 31 ? 59 73 ? 101

b. 134 111 ? 65 ? 19

c. ? 84 ? 58 45 32 19 ?

d. $0 \quad \frac{1}{4} \quad \frac{1}{2} \quad \frac{?}{?} \quad 1 \quad 1\frac{1}{4} \quad ?\frac{?}{?} \quad 1\frac{3}{4} \quad 2$

Extension

A student starts a bank account with $30 right after New Year's Day. The student also earns $12 every week, and starts putting that money in the bank the week following New Year's Day. How many weeks after New Year's Day will the student have saved the $198 necessary for a stereo system?

Activity 9: Common Divisors and Multiples

(Transparency 14)

This is another problem that would be extremely tedious using paper and pencil but that is relatively easy to test using the calculator. The calculator allows users to try as many values as they need to find the greatest common divisor and lowest common multiple of each pair of numbers. The problem is a case of straightforward procedure: (1) conjecture, (2) test via computation, and (3) return to step 1, until a tentative solution is found. These common types of problems are prime candidates for calculator solution.

Objective

Given a calculator, learners will determine the greatest common factor (GCF) and the lowest common multiple (LCM) of a pair of numbers.

Solution Procedure

Learners should proceed here via estimation and calculator verification. The calculator affords them the luxury of rapid confirmation or rejection of estimates. For instance, to find the GCF in (b), learners could reasonably attempt these possibilities:

Try 5:	30 [÷] 5 [=] (6 [ok])	45 [÷] 5 [=] (9 [ok])
Try 10:	30 [÷] 10 [=] (3 [ok])	45 [÷] 10 [=] (4.5 [no])
Try 15:	30 [÷] 15 [=] (2 [ok])	45 [÷] 15 [=] (3 [ok])

Answers

a. 3 **b.** 15 **c.** 6 **d.** 11

Solution to Extension Problem

The extension problem is solved as the main problem was, except that learners should begin with the smallest candidate numbers that might be LCMs of each pair. For instance, (a) could reasonably be solved via the following reasoning and calculation:

Try 9: 9 [÷] 9 [=] (1 [ok]) 9 [÷] 6 [=] (1.5 [no])

Try 12: 12 [÷] 9 [=] (1.3333333 [no])

Try 18: 18 [÷] 9 [=] (2 [ok]) 18 [÷] 6 [=] (3 [ok])

The solutions are:

a. 18 **b.** 90 **c.** 36 **d.** 110

Common Divisors and Multiples

Below are several pairs of numbers. For each pair, figure out the biggest number that divides evenly into both numbers.

a. 9 , 6 **c.** 18 , 12

b. 30 , 45 **d.** 11 , 110

Extension

For each pair, figure out the smallest number that both numbers divide into evenly.

Activity 10: Multiplication and Division Patterns

(Transparency 15)
This problem is analogous to the Arithmetic Sequences problem. Learners should use the [Cons] key of the calculator to regenerate the geometric progression problems. After estimating a potential common ratio, learners can use the calculator to see if that common ratio generates the subsequent values in the progression. If not, they can revise their estimate and verify again. The necessary keystrokes to generate an entire progression are detailed in the Solution Procedure for this problem. The Solution to the Extension Problem is given on the next page.

Objective

Given a calculator, learners will determine terms in geometric sequences.

Solution Procedure

Some learners may solve these problems by mental arithmetic, but they should verify their solutions on the calculator. Most learners will proceed by estimation and trial and error via the calculator to discern the common ratio. Once an estimated common ratio is derived, learners can employ the [Cons] key on the calculator to go through the progression step by step. In (a) learners could try a common ratio of 3 and could generate the entire progression by pressing the following keystrokes:

[×] 3 [Cons] 2 [Cons] [Cons] [Cons] [Cons] [Cons] [Cons]

At that point, they could simply fill in the missing terms in the answer.

Answers

a.	2	6	18	**54**	162	486	**1458**
b.	1200	240	48	9.6	**1.92**	0.384	**0 .0768**
c.	5	$\frac{5}{3}$	$\frac{5}{9}$	$\frac{5}{27}$	$\frac{5}{81}$	$\frac{5}{243}$	

Solution to Extension Problem

The extension problem is a variation of the famous "corn kernels on a chessboard" math problem. The secret is the rapid growth of a geometric progression, even when the common ratio is two. Learners can use the [Cons] key and the constant operation of multiplication by 2. They will need to write the terms of the generated progression. When the iteration counter on the display shows 13 (since the progression is initiated with one, 13 doublings produce the amount for the 14th day), the display will show 8192. Learners may then sum the daily amounts they've written down to arrive at the total $16,383. There are a couple of other ways to approach this problem, including using the memory keys in conjunction with the [Cons] key to sum the running total in memory and using the [y^x] key. The solution suggested is the most straightforward.

Multiplication and Division Patterns

Each row of numbers below is a geometric sequence (formed by consecutive multiplication or division by the same number). Fill in the missing numbers so that they match the pattern in that row. Hint for (c.): Multiply by a fraction.

a.	2	6	18	?	162	486	?
b.	1200	240	48	9.6	?	.384	?
c.	5	$\frac{5}{3}$	$\frac{5}{9}$	$\frac{?}{?}$	$\frac{5}{81}$	$\frac{?}{?}$	

Extension

Suppose I give you $1 today. Tomorrow I give you twice as much ($2). Every day thereafter, I give you twice as much as the day before. How much (total) will you have after 2 weeks? (Count the $1 day as the first day.)

9

Workshop: Using a Calculator to Develop Higher-Order Thinking Skills Grades 7-9

Instructions to Presenters

Overview

This workshop is designed to illuminate the ways in which calculators aid problem solvers faced with higher-order math problems. The problem-solving activities included here demonstrate the near necessity of a calculator to:

- Free the problem solver of the burden of manual computation so that he or she may concentrate on higher-order strategies.
- Allow experimentation with varying values to discover patterns and solutions.

There are five fully developed problems (Activities 1-5) and five supplementary problems (Activities 6-10) in this 7-9 workshop. It may be advisable to intersperse details from the front material as necessary. The quotations from the National Council of Teachers of Mathematics (on Transparency 1 in Chapter 6) provide participants with a powerful viewpoint on the importance of calculators and higher-order thinking. All of the quotes refer to the development of a mathematics curriculum. Another way to fill out the presentation would be to have participants create their own problems, or to open the presentation for discussion.

Activity 1: Maximize the Resulting Fraction

(Transparencies 1 and 2)
This problem asks learners to experiment with the calculator to determine how to maximize fractional computations involving the four arithmetic operators (addition, subtraction, multiplication, and division). If learners experience difficulties, it will probably be with the subtraction and division problems. In those cases, the second maximizing fraction is the smallest composable fraction $\frac{2}{9}$. The special capabilities of the Math Explorer allow for simple computational experimentation with fractions. The problem can be altered so that the learners may not use any digit more than once. Thus, after the 2 and the 9 are used once, learners must choose other digits for the other fraction in each problem. The Extension Problem provides the opportunity to make an important generalization about division.

Mathematical Content

Combining fractions, arithmetic operations, maximization, calculator skills

Higher-Order Thinking Skills

Reason about phenomena, develop conjectures, build abstractions, solve problems, validate assertions

Potential Problem-Solving Techniques

Identify what is being asked, experiment, record the data, form tentative solutions, verify the solution

Calculator Notes

Teachers need to be aware of the special fractional capabilities of the Math Explorer calculator. Refer to Chapter 2 for detailed instructions for using these advanced capabilities.

Solution to Extension Problem

The answer to both questions is yes. The Extension Problem is again approached by intuition and computational trials. Although learners cannot test all possible values, they should be able to generalize the truth of the two assertions by experimenting with successive calculator divisions.

Maximize the Resulting Fraction

Objective

Fill in the boxes below with numbers between 2 and 9 so that the operations on the fractions produce the LARGEST possible results.

a. $\frac{\square}{\square} + \frac{\square}{\square} = \text{max?}$

c. $\frac{\square}{\square} - \frac{\square}{\square} = \text{max?}$

b. $\frac{\square}{\square} \times \frac{\square}{\square} = \text{max?}$

d. $\frac{\square}{\square} - \frac{\square}{\square} = \text{max?}$

Extension

Is the result you obtained by dividing a number by a value smaller than 1 always larger than the original number? Is the result you obtain by dividing a number by a value larger than 1 always smaller than the original number?

Maximize the Resulting Fraction

Objective

Given a calculator, learners will determine where to place the digits 2 through 9 so that the results of arithmetic operations with fractions are maximized.

Solution Procedure

1. Start with an estimate. For instance, the result of a multiplication is usually larger if the two numbers being multiplied are large. Try these values on the calculator:

   5 [/] 3 [×] 9 [/] 3 [=] [Simp] [=] [Simp] [=] ($\frac{5}{1}$)

   6 [/] 2 [×] 9 [/] 2 [=] [Simp] [=] ($\frac{27}{2}$)

2. Record the results of the computation in a table of values.

3. Try other values until a pattern of "larger" results emerges.

   9 [/] 2 [×] 9 [/] 2 [=] ($\frac{81}{4}$ [the largest value])

4. Immediately or eventually, learners will observe the pattern of results for each fractional operation.

Learners can use the [F⇄D] key to convert the fractions to decimals for comparison.

For instance, the answer $\frac{81}{4}$ can be converted easily:

9 [/] 2 [×] 9 [/] 2 [=] [F⇄D] (20.25)

Answers

a. $\frac{9}{2} + \frac{9}{2}$

b. $\frac{9}{2} \times \frac{9}{2}$

c. $\frac{9}{2} - \frac{2}{9}$

d. $\frac{9}{2} \div \frac{2}{9}$

Activity 2: Sally's Share of Pay

(Transparencies 3 and 4)
This problem exploits the power of the Math Explorer Integer Division key. Using this key, the problem proceeds simply, as detailed in the Solution procedure. To solve this problem without access to an integer division key involves one of several time-consuming algorithms. Learners could use the standard division key, and then convert the decimal part of the division result to an integer by constructing a proportion between the decimal fraction and the "base" number of kids. An alternative is performing a standard division, multiplying the integer part of the division by the number of kids, and then subtracting that result from the total of dollars earned. This would yield Sally's share. It should be clear that the integer division operation provides by far the simplest solution to this problem. The Extension problem provides the opportunity for mathematical generalization.

Mathematical Content

Integer division, calculator skills

Higher-Order Thinking Skills

Reason about phenomena, develop conjectures, build abstractions, solve problems, validate assertions

Potential Problem-Solving Techniques

Understand the problem setting, identify what is being asked, record the data, make a chart, verify the solution

Calculator Notes

Teachers should make learners aware of the special integer division capabilities of the Math Explorer calculator. Refer to Chapter 2 for detailed instructions for using these advanced capabilities.

Solution to Extension Problem

The most Sally can earn from any group of N kids is equal to the number of kids in the group minus one (N - 1 dollars). With a group of 8 kids she could earn at most $7, with a group of 12 kids, $11. The Extension problem can be approached via calculator experimentation with various sets of values until the pattern of remainders is apparent. A good approach would be to increase the dollar amount by a dollar at a time and observe the remainder pattern for different divisors.

Sally's Share of Pay

Sally gives a bit of help to several groups of kids who earn money. Because she doesn't work much, she asks only that the kids give her whatever is left over after they divide up their earnings equally among themselves (in whole dollar amounts). For instance, if five kids earned $52, each of them would get $10, and Sally would get the remaining $2. Figure out how much money Sally earned in the week shown below. The groups divided their shares evenly each day (to the nearest whole dollar) and gave her what was left over:

Monday	8 kids divided $70
Tuesday	9 kids divided $85
Wednesday	14 kids divided $165
Thursday	12 kids divided $140
Friday	11 kids divided $120

Extension

Eight kids work on a job. They divide their earnings into equal whole-dollar shares and give the rest to Sally. What is the most number of dollars that Sally could receive? What is the most if 12 kids work on a job? Can Sally always know the most she could make if she knows how many kids plan to work on a job?

Sally's Share of Pay

Objective:

Given a calculator, learners will compute and sum the remainders of divisions.

Solution Procedure:

1. Use the integer division key on the calculator to find the remainder of Monday's pay after the other kids have taken their share.

   70 [INT÷] 8 [=] (Q 8, R 6)

2. Write down the remainder (6).
3. Repeat the procedure for the other four days.

   85 [INT÷] 9 [=] (Q 9, R 4)

   165 [INT÷] 14 [=] (Q 11, R 11)

   140 [INT÷] 12 [=] (Q 11, R 8)

   120 [INT÷] 11 [=] (Q 10, R 10)

4. Sum the remainders to find Sally's share of pay for the week.

   6 + 4 + 11 + 8 + 10 = (39)

5. There are other ways to use the calculator to solve this problem, but all are far more difficult than using the [INT÷] key.

Answer

Sally's share for the week is $39.

Activity 3: What Must the Units Be?

(Transparencies 5 and 6)
This is a relatively easy problem. The trick is to divide the two numbers in each problem in such a way as to produce a recognizable conversion factor. If a division operation is performed in "both directions" with the two numbers in each operation, a recognizable conversion factor is revealed. The applicable conversions are fairly simple, but some learners may need a table of conversion factors. The Extension problem involves a sequence of conversions to go from one unit to another. The details of the most appropriate conversion are presented in the Extension problem solution.

Mathematical Content

Measurement unit conversion, arithmetic operations, calculator skills

Higher-Order Thinking Skills

Reason about phenomena, develop conjectures, solve problems, validate assertions

Potential Problem-Solving Techniques

Identify what is being asked, estimate, experiment, form tentative solutions, verify the solution

Solution to Extension Problem

One mile is farther than 1500 meters (both are race distances for track and field events). Fifteen hundred meters equal approximately 4950 feet (a mile is 5280 feet). The extension problem is a matter of determining a "conversion path" from one of the units to the other. Learners will probably have more success working from meters to miles than from miles to meters. One possible "path" goes from meters to inches to feet, which can then be compared to the commonly known 5280 feet in a mile.

1500 [×] 39.5 [=] [÷] 12 [=] (4937.5 [feet])

What Must the Units Be?

Below are some "equal measurements." Decide in what units the second measurement must be expressed to make the equality true. (The units may be English or metric.)

144 inches = 12 __________

6 yards = 216 _________

15000 grams = 15 __________

7 miles = 12,320 ______

Extension

In the Olympics there are races of 1 mile and 1500 meters. Which is farther?

What Must the Units Be?

Objective

Given a calculator, learners will determine the units involved in given measurement conversions.

Solution Procedure

1. Learners need to know or have access to basic conversion factors (such as inches per yard).
2. Divide the two numbers to find the conversion factor. For instance, in the second problem, dividing 216 by 6 on the calculator yields 36, the number of inches in a yard.

   216 [÷] 6 [=] (36)

3. Proceed in a similar manner to find the conversion factors and the appropriate units for each of the problems on the page. It may be necessary to perform the division in both directions before a conversion factor is recognized. For instance, in the second problem:

   6 [÷] 216 [=] (0.0277778 [not a known conversion factor])

However,

216 [÷] 6 [=] (36 [inches per yard])

Answers

a. 144 inches = 12 feet

b. 6 yards = 216 inches

c. 15 000 grams = 15 kilograms

d. 7 miles = 12,320 yards

Activity 4: Missing Fractions

(Transparencies 7 and 8)
This set of fractional calculations can be solved via calculator experimentation or by inverse operations (solving the equations as one would solve them in algebra). Again, the fractional prowess of the Math Explorer allows learners to investigate these problems using iterative exploration if they wish. By plugging possible solutions into the calculation, learners can zero in on the correct result. If learners use the iterative approach (as opposed to equation solving), they should record the results of their successive attempts in a table so that a pattern can be discerned.

Mathematical Content

Fractional arithmetic, inverse operations, estimation, calculator skills

Higher-Order Thinking Skills

Reason about phenomena, develop conjectures, solve problems, build abstractions, validate assertions

Potential Problem-Solving Techniques

Identify what is being asked, estimate, experiment, record the data, form tentative solutions, verify the solution

Calculator Notes

Teachers need to be aware of the special fractional capabilities of the Math Explorer calculator. Refer to Chapter 2 for detailed instructions for using these advanced capabilities.

Solution to Extension Problem

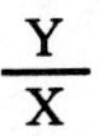

The extension problem may allow learners to express the multiplicative inverse in symbolic terms. They may experiment with several values on the calculator to verify that inverting any fraction and multiplying the inverted fraction times the original fraction yields 1, but they may or may not be able to relate that to symbolic notation.

Missing Fractions

The computations below involve fractions. However, there are boxes where some fractions should be. See if you can figure out what fractions must be placed in the boxes to make the computations true.

a. $\frac{1}{3} + \frac{\square}{\square} = \frac{2}{3}$

b. $\frac{1}{\square} \times \frac{2}{3} = \frac{1}{6}$

c. $\frac{2}{3} - \frac{1}{\square} = \frac{1}{2}$

d. $\frac{4}{5} \div \frac{\square}{\square} = 1\frac{3}{5}$

e. $\frac{2}{3} \times \frac{\square}{\square} = 1$

Extension

Every fraction has a "multiplicative inverse." The multiplicative inverse of any fraction is the fraction that it must be multiplied by to make 1.

If we had some fraction, any fraction — let's call it $\frac{X}{Y}$, how could we represent its multiplicative inverse? In other words:

$$\frac{X}{Y} \times \frac{\square}{\square} = 1$$

Missing Fractions

Objective

Given a calculator, learners will determine missing fractions in arithmetic operations involving fractions.

Solution Procedure

1. Make an estimate of what fraction would combine with the given fraction to produce the indicated result. For instance, in (b), a reasonable first estimate would be $\frac{1}{2}$. Try that value using the calculator:

   1 [/] 2 [×] 2 [/] 3 [=] [Simp] [=] $(\frac{1}{3})$

2. That result is too large, so try a smaller fraction, $\frac{1}{3}$:

   1 [/] 3 [×] 2 [/] 3 [=] $(\frac{2}{9})$

3. Closer. Keep trying until the appropriate fraction $(\frac{1}{4})$ is found.

   1 [/] 4 [×] 2 [/] 3 [=] [Simp] [=] $(\frac{1}{6})$

4. The problems may be solved directly through the use of inverse operations to "solve each equation." Also, some learners' insight will allow them to limit themselves to denominators that are factors or multiples of the given denominators.

Answers

a. $\frac{1}{3} + \frac{1}{3} = \frac{2}{3}$

b. $\frac{1}{4} \times \frac{2}{3} = \frac{1}{6}$

c. $\frac{2}{3} - \frac{1}{6} = \frac{1}{2}$

d. $\frac{4}{5} \div \frac{1}{2} = 1\frac{3}{5}$

e. $\frac{2}{3} \times \frac{3}{2} = 1$

Activity 5: Negative Targets

(Transparencies 9 and 10)
This is a straightforward problem set. It can be approached in the same way as the previous problem — by successive approximation and calculator verification or by equation solving using inverse operations. Learners can either solve the problems directly by "solving for the empty box" as in algebra, or by estimating the value of the empty box and then verifying their estimates using the calculator. They should progress toward a solution by iteration and successive zeroing in.

Mathematical Content

Arithmetic operations, inverse operations, negative numbers, estimation, calculator skills

Higher-Order Thinking Skills

Reason about phenomena, develop conjectures, solve problems, build abstractions, validate assertions

Potential Problem-Solving Techniques

Identify what is being asked, estimate, experiment, record the data, form tentative solutions, verify the solution

Solution to Extension Problem

The answer is - X . The extension problem may allow learners to express the additive inverse in symbolic terms. They may experiment with several values on the calculator to verify that the negative of any number added to the number yields 0, but they may or may not be able to relate that to symbolic notation.

Negative Targets

The computations below have boxes where numbers should be. Use the calculator and see how quickly you can figure out what number must be placed in each box to make each computation true. The result of every computation is negative. The missing numbers could be positive or negative.

a. $73 - \square = -51$

b. $118 \times \square = -7670$

c. $-36 \div \square = -72$

d. $170 + \square = -170$

Extension

Every number has an additive inverse. The additive inverse of any number is the number that must be added to it to make 0. If we had some number, any number — let's call it X (X the unknown) — how could we represent the additive inverse of X? In other words:

$X + \square = 0$

Negative Targets

Objective

Given a calculator, learners will determine missing terms in computations that produce negative results.

Solution Procedure

1. Estimate what the missing number must be, and perform a test computation on the calculator. For instance in (a), estimate 100 (the number must be greater than 73, because the result of the subtraction is negative:

   73 [−] 100 [=] (-27 [too big, subtract more])

   73 [−] 120 [=] (-47 [very close, try 4 larger])

   73 [−] 124 [=] (-51 [Correct!])

2. Eventually, learners will zero in on all the missing numbers.

3. Of, course, the problems may be solved directly through the use of inverse operations to solve each equation.

Answers

a. 73 - **124** = -51

b. 118 x **-65** = -7670

c. -36 ÷ **0.5** = -72

d. 170 + **-340** = -170

Activity 6:
Two Shapes and Their Areas

(Transparency 11)

This activity requires learners to solve several problems. The first problem is not complicated. They must compute the areas of a square and circle, given the side and diameter, respectively. This is easily performed using the calculator. The [π] key allows that constant to be entered accurately and without memorization. The second part of the main problem is more difficult. Here, learners may use the calculator as an environment for approaching the desired value of the circle's diameter as described in the Solution Procedure for that problem, or they must solve the equation of the area of a circle for the required radius (r), and double that to discover what the diameter must be. The Solution to the Extension Problem is given on the next page.

Objective

Given a calculator, learners will determine which shape, a square or a circle, encloses the greatest area for a given perimeter.

Solution Procedure

Learners should draw a diagram of a square and a circle. They should use the calculator to compute the area of each and write it down. To answer part two, learners will need to do one of two things: Either they must use the equation for the area of a circle, fill in the computed area of the square as A, solve for r, and then double that to ascertain the diameter, or they must proceed by trial and error on the calculator to increase the diameter of the circle until its area approaches the area of the square. In either case, the [π] key of the calculator will speed computation by automatically entering the value of pi. To match the square's area (25,600), learners might start by trying a circle with diameter 170:

170 [÷] 2 [=] [x^2] [×] [π] [=] (22698.007 [too small, try 180])

180 [÷] 2 [=] [x^2] [×] [π] [=] (25446.9 [very close, try again])

Answer

The square has the greater area. The diameter of the circle would need to be 181 centimeters to yield the same diameter as the square.

Solution to Extension Problem

The circle is more efficient. For instance, in the example above, the perimeter of the square is 640, whereas the circumference of the circle to enclose the same area is only 568. The extension problem can be resolved by using two shapes with equal areas (as in the above problem if successfully completed) and working backward from the areas, to the dimensions, to the perimeters. In the example above, for instance, equal areas are produced by a square with a side of 160 and a circle with a diameter of 181. A square with a side of 160 has a perimeter of 640 ($P = 4s$). A circle with a diameter of 181 has a circumference of 568 ($C = \pi d$). Learners unaware of the appropriate formulas or how to apply them may resort to successive approximation via trial and error to arrive at a solution, although this may prove too frustrating an approach.

Two Shapes and Their Areas

The side of a square measures 160 cm. The diameter of a circle measures 160 cm. Which of the two shapes has the greater area? How long would the diameter of the circle have to be so that the area of the circle was the same as the area of the square? (Answer to nearest whole number.)

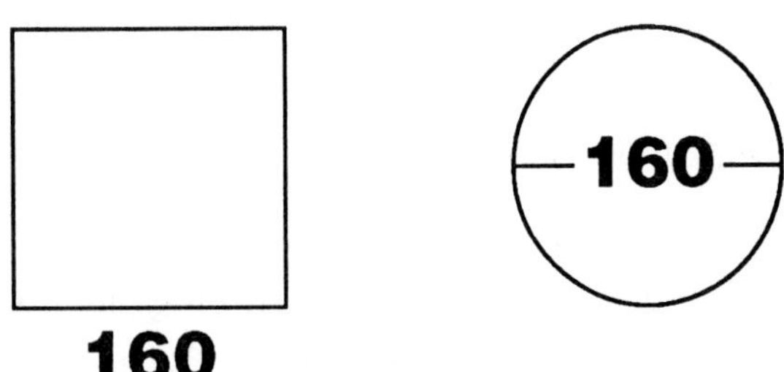

Extension

Which shape is more efficient at enclosing space, a square or a circle? (In other words, which shape requires the smallest perimeter to enclose the same amount of area?)

Activity 7: Multiplication and Division Patterns

(Transparency 12)
This problem contains no tricks. Learners should use the [Cons] key of the calculator to regenerate the geometric progression problems. After estimating a potential common ratio, learners can use the calculator to see whether that common ratio generates the subsequent values in the progression. If not, they can revise their estimate and verify again. The necessary keystrokes to generate an entire progression are detailed in the Solution Procedure for this problem.

Objective

Given a calculator, learners will determine terms in geometric sequences.

Solution Procedure

Some learners may solve these problems by mental arithmetic, but they should verify their solutions on the calculator. Most learners will proceed by estimation and trial and error via the calculator to discern the common ratio. Once an estimated common ratio is derived, learners can employ the [Cons] key on the calculator to go through the progression step by step. In (a), learners could try a common ratio of 3 and could generate the entire progression by pressing the following keystrokes:

[×] 3 [Cons] 2 [Cons] [Cons] [Cons] [Cons] [Cons] [Cons]

At that point, they could simply fill in the missing terms in the answer.

Answers

a.	2	6	18	**54**	162	486	**1458**
b.	1200	240	48	9.6	**1.92**	0.384	**0.0768**
c.	6	-30	150	**-750**	3750	-18,750	**93,750**
d.	120	-30	7.5	-1.875	**0.46875**	**-0.1171875**	0.0292969

Solution to Extension Problem

There are only two cases where a zero term may appear in a geometric progression (and they are trivial): (1) The common ratio is 0 (whatever the first term is, all the remaining terms become 0). (2) The first term is 0 (whatever the common ratio is, every term is 0). The answer to the second question is that the signs will always alternate, except in the first trivial case, when the initial term is negative and the common ratio is 0. The Extension questions can be answered by mathematical insight or by experimentation with different sequences. Learners will need to be aware of the definition of a geometric sequence to realize that the trivial cases involving 0 are definitionally allowed. Most learners should agree, via experimentation and insight, that the signs of a geometric sequence with at least one negative term must alternate — which is accurate (with the trivial exception just noted).

Multiplication and Division Patterns

Each row of numbers below is a geometric sequence (formed by consecutive multiplication or division by the same number). Fill in the missing numbers so that they match the pattern in that row.

a.	2	6	18	?	162	486	?
b.	1200	240	48	9.6	?	0.384	?
c.	6	-30	150	?	3750	-18,750	?
d.	120	-30	7.5	-1.875	?	?	0.0292969

Extension

Is it possible to have a zero (0) as a term in a geometric sequence? If a standard geometric sequence contains both positive and negative numbers, will the two signs always alternate term by term?

Activity 8: Indivisible Dollars

(Transparency 13)
This problem is an introduction to prime numbers. Learners must investigate whether given values (20, 50, and 100) are prime, and if they are not, learners must determine the next lowest prime below each. This involves exploring divisibility (or factorability) using the calculator. Without a calculator this would be an extremely time-consuming and valueless set of problems. Learners should proceed by going one lower from the original number until they find a prime. To check whether each successively smaller number is prime, they should divide the number by small factors, such as 2, 3, 5, and so forth, until they verify to themselves that they have found a number that has no factors (except 1 and the number itself).

Objective

Given a calculator, learners will determine the largest prime number less than or equal to a given number.

Solution Procedure

Assuming that learners do not have access to a table of prime numbers, the simplest procedure is to begin with the original amount (20, 50, or 100), and experiment with the calculator to see if it is factorable. It would be best for learners to work downward from the original number, checking each next smaller integer amount for divisibility, until a tentative prime number is found. The best strategy may be to begin with the most common potential prime factors and work upward (2, 3, 5, 7, etc.). To test 20 in (a):

20 [÷] 2 [=] (10 [so 20 is not prime; try 19, etc.])

Answers:

a. $20 — should ask for **$19.**

b. $50 — should ask for **$47.**

c. $100 — should ask for **$97.**

Solution to Extension Problem

The factors of 100 are: 1, 2, 4, 5, 10, 20, 25, 50, 100. The extension problem is a matter of calculator experimentation (via division operations) and/or of mathematical insight (for instance, that since 20 is a factor, so must be 2, 10, 5, and 4).

Calculator Notes

Teachers may wish to make students aware of how to use either the memory storage and memory recall keys or the "register swap" key on the Math Explorer calculator. (The swap key is the [x⇄y] key.) Either the memory recall or register swap key can be used quickly to retrieve a potential dollar amount request (20, 19, etc.) after a division operation has been used to determine whether a number is a factor of that amount. This saves rekeying the original amount being tested after each factoring attempt. The swap key accommodates this capability in fewer keystrokes. Refer to Chapter 2 for detailed instructions for using these advanced capabilities.

Indivisible Dollars

Jake's grandparents like to give him and his four brothers money at holidays. Grandpa always tells Jake the maximum amount of money he is willing to give him. Jake can have that amount, or any lesser amount (all amounts are in even dollars). If the amount Jake selects can be evenly divided by any whole number (except the number itself and the number 1), then Jake must divide the money with his brothers. If Grandpa offered Jake $10, it would be best to ask for $7, because 10 is divisible by 2 and 5, 9 is divisible by 3, and 8 is divisible by 2 and 4 (whereas the only factors of 7 are 7 and 1). How much should Jake ask for to retain the most from each of the offers below?

a. $ 20 offer

b. $ 50 offer

c. $ 100 offer

Extension

Make a list of every number that is a factor of (divides evenly into) 100.

Activity 9: A Rectangular Garden

(Transparency 14)
This problem asks learners to discover the rectangular shape that encloses the most area for a given perimeter — a square. Some learners may know in advance that a square is the most efficient area encompasser. Most learners will not and will proceed by trying different shapes, recording the resulting areas, and forming a generalization (that the more square the shape, the greater the area). It is helpful to draw a diagram, calculate the respective areas using the calculator, and reach a conclusion. It should be pointed out to the participants that without a calculator, this problem would get bogged down in computation (simple multiplication), which has no educational benefit. With a calculator, learners may concentrate on strategy and patterns.

Objective

Given a calculator, learners will discover the rectangular shape that encompasses the greatest area for any given perimeter.

Solution Procedure

1. Draw a diagram of any rectangle whose perimeter totals 180, such as:

   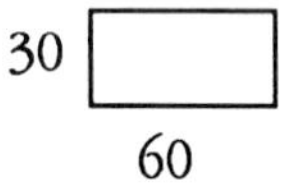

2. Use the calculator to compute the area, and write it down.

   30 [×] 60 [=] (1800)

3. Vary the shape (and consequently the dimensions) of the rectangle, calculate the area, and write it down.
4. Learners should eventually hypothesize and conclude that the closer the rectangle is to being square, the greater the area.

Answer

The greatest area is 2025 square meters (a 45 × 45 square). Without a calculator, this result would need to be arrived at either by insight or via consecutive paper-and-pencil or mental calculations.

Solution to Extension Problem

A square will always enclose the greatest area for any given perimeter. The extension problem is resolved by experimenting with rectangles of several perimeters to ensure that the square of that perimeter yields the maximum area.

A Rectangular Garden

You have 180 meters of fence. You want to use it to enclose a rectangular-shaped garden. What is the largest area that you can enclose with the fence?

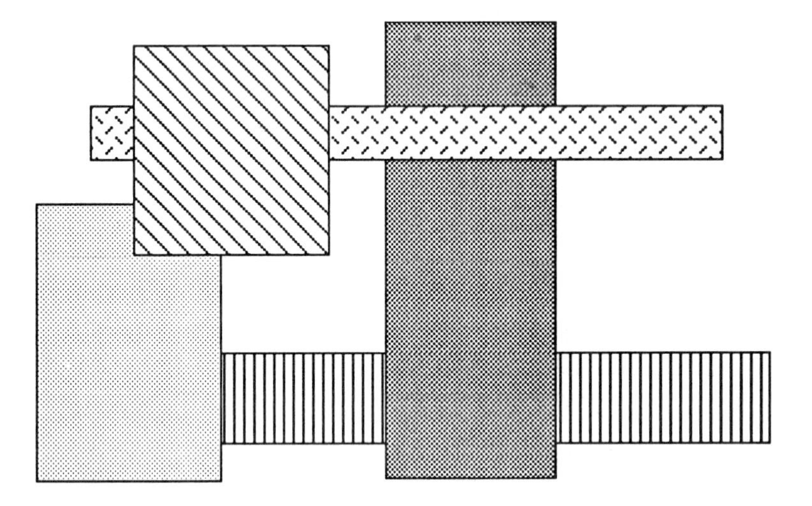

Extension

Is there always a rectangle of maximum area for any given perimeter? If so, what can be said about it?

Activity 10: Selling Maple Syrup

(Transparency 15)
This problem has at least three solutions. It is possible that participants may discover another. The three solutions are listed in the Solution Procedure. This problem is approached by calculator experimentation. Learners should be allowed time to try multiples of factors to determine whether a combination of these multiples can add to 100. The Extension problem asks learners if a 50-liter order for maple syrup can be met. Learners who have come upon a solution that has all even numbers of factors (2 or 4 or 6 of each factor) for the 100-liter order will believe that an order for 50 liters can be met (by halving the counts they discovered). Learners whose solution to the 100-liter problem has an odd count for any factor (1, 3, 5, etc.) may decide that the 50-liter order cannot be met, since they cannot halve their solution.

Objective

Given a calculator, learners will determine how to allocate volumetric containers to reach a specified final amount.

Solution Procedure

1. Start with one of the largest factors, say 18. Estimate that it could go into 100 five times. Verify that with the calculator.
2. Estimate whether any multiple of a smaller factor could "fill" the remaining "distance" to 100 (i.e., 10). Because 9 is the only small-enough possibility and it doesn't work, go back and start over.
3. See whether 4 times 18 might be a good starting point.
4. Estimate whether any multiples of a smaller factor could "fill" to 100.

   72 [+] 1 [×] 15 [+] 1 [×] 13 [=] (100) Yes

Answer

Two other solutions proceed from calculator experimentation:

4 × 15 liters (= 60) + 1 × 18 liters (= 18) +

1 × 13 liters (= 13) + 1 × 9 liters (= 9) = 100 Total

2 × 15 liters (= 30) + 4 × 13 liters (= 52) +

1 × 18 or 2 × 9 (= 18) = 100 Total

Solution to Extension Problem

Any multiple of 100 liters can be satisfied by using the same container sizes and multiplying their count by the ratio of the multiple of 100 to 100. The 50-liter order can be met by halving the second solution (1 × 15, 2 × 13, 1 × 9) but may seem impossible to learners who have found a solution to the 100-liter problem whose counts cannot be halved. Twenty-five liters cannot be met.

Selling Maple Syrup

The Acme Maple Syrup Company wholesales maple syrup in large cans and barrels. The containers measure 9, 13, 15, and 18 liters only. The company must meet each customer's order exactly, or lose the sale.

How can an order for 100 liters of maple syrup be met?

Extension

Because an order for 100 liters can be met, does that mean that orders for 200, 300, 400 and other multiples of 100 liters can also be met? Can an order for 50 liters be met? Twenty-five liters?

10

Interdisciplinary Calculator Projects

In the present era of strong recommendations for integrating the curriculum, calculators can be invaluable for accomplishing this endeavor. This chapter is designed to make the use of the calculator possible in interdisciplinary projects. First, there are two projects for each grade level—primary (K-3), intermediate (4-6), and upper grades (7-9)—outlined in detail. These projects utilize calculators and various disciplines. Each is described in detail. The second part of the chapter gives specific activities related to geography, science, ecology, history, and consumerism. Finally, the last section lists interdisciplinary calculator application ideas. For these ideas the teacher is required to provide the problem and the plan for using and solving the problem in the classroom. It is hoped that teachers can find several ideas applicable to their grades and classroom expectations.

Interdisciplinary Calculator Projects

These projects are designed as class or small-group projects. Modify projects to fit time restrictions, class competence, and availability of the situations.

Primary

Car Counting. Take your students to a quiet street near the school. Assign each child a common car color. Have the children use calculators to count the number of cars of their color that pass by during a 10- to 15-minute interval. When you return to the classroom, help the children make a picture chart or line graph to show the results.

The next day at the same time, take the children to a busy street in the area. Ask them to predict the number of cars of each color that will pass by, using the graph. Record their estimates. Then have the children again count the number of cars of each color. Use these numbers to make another graph. Have the students compare the two graphs and note how close their predictions came to the actual numbers.

Repeat the activity on two other days but at other times. Compare the two graphs from this session, then compare them with the first two. Discuss the results and some of the possible reasons the traffic counts changed or stayed the same.

Planning and Planting a Garden. Select a small plot of land near the school where you can plant a garden, or use an empty wooden, metal, or plastic box filled with soil. Or, lay out a "plot of land" on the chalkboard, drawing a sketch of the garden space.

Decide with the children what vegetables to plant. After determining the space needed between rows, use a calculator to determine how many rows can be planted in that area. Using the information on the seed packages, figure out how many seeds of each variety you should plant and calculate the total.

Then plant the garden. Record on a picture graph the amount of crop each plant produced or its growth rate.

Intermediates

Population Growth. Have a group of students visit your town or city hall to find out the community's total population for each census taken during the last 50 years. Then graph the results in class. Ask the students to use calculators to compute the amount of increase or decrease between each census period, and graph those results.

Using calculators, set up a graph showing school and district enrollment figures for as many years as are available. Ask your students to find what percent the school population is of the community population.

Help your students use calculators to predict future population growth for the community and school district, based on these facts and figures. Discuss reasons for increases and/or decreases in the school and community populations. Will there be a need for more schools, post offices, police departments, and so on?

Organizing a Recycling Program. Have a student committee contact your community's public works department to request information about the organization and costs of the local recycling program. Then suggest that students consider organizing a similar school program. They will need to know what materials are gathered, how and where they are stored, and how they are transported to the recycling center.

The students will have to decide whether to collect newspapers, glass bottles, aluminum cans, or all three. Using calculators, they can compute the costs involved in providing a receptacle and in transporting materials to the center. Then they can determine how many pounds of each type of material they would need to collect to reach a breakeven point. You may want to seek school and community support for the project.

Upper Grades

School Election Predictions and Polls. Ask the students to research polling techniques and procedures. Work with them to set up a sample procedure for predicting student elections in your school. First, they will have to find out the number of students in each class by grade level. Then, using calculators, have them compute the number of children in each class to be counted in the sample.

Students will then survey the random sample and use their projections to predict election results.

Have the class compare the actual results with their predictions and compute the differences using calculators. In discussing the comparisons and possible reasons for the variations, try to set up improved sample procedures for future elections. Students might want to apply their findings to setting up community election sampling requirements and go on to poll and predict outcomes.

Pollution Analysis. Have a group of students contact your state, regional, or county department of environmental protection and ask for statistics on air pollution for your community for the past 10 to 15 years. If such figures are not available, contact a nearby large city. If possible, obtain statistics on both air and water pollution from various causes, including automobiles and industry.

Using calculators, students can compute the differences for particular types and causes of pollution in each 2- or 3-year period. Organize the results into a line graph. On the basis of the graph's information, have the class predict future pollution rates and discuss precautions that can be taken.

Typical Calculator Interdisciplinary Activities

The most typical calculator uses involve ways to extend the curriculum. Each computation grows out of and enriches either a geography, science, ecology, history, or consumerism lesson. Each computation also represents a mathematics application and discovery that might not be possible without the calculator because of time or ability levels.

1. Compute and compare the distances from Los Angeles to New York via a northern route and a southern route.
2. Compute the approximate number of beverage bottles in one aisle of a grocery store by multiplying the number of bottles across each row by the number of rows deep and the number of shelves in each row.
3. Compute the average speed of a car, given the time and distance traveled.
4. Using current exchange rates, determine the value of $10 in the currency of several countries.
5. Compute the average amount of rainfall in your area for a month, the average temperature for a week, and the average wind velocity for 10 days.
6. Compute what percentage of your state's population your city or town's population represents.
7. Find the average temperature in the United States on a particular day, using the temperatures of all 50 state capitals.
8. Round the populations of the major cities of your state to the nearest thousand and add them. Then add the actual populations. Compare the sums and compute the percent of error.

9. Compare the average cost of an automobile in 10-year intervals from the early part of this century to the present. Graph the results.
10. Determine the gear ratios of various bicycles, comparing wheel radius to sprocket wheel radius.
11. Compare land surface area in your state to water surface area and determine what percentage is water surface area.
12. Compute the average length of term of office of U.S. presidents.
13. Compute the number of calories you had for breakfast, lunch, and dinner. Then determine the total for the day and week and your average daily calorie intake.
14. Find the height of your school building. First determine the ratio of the length of your shadow to your actual height; then measure the height of the shadow cast by the school and use the ratio to estimate the building's height.
15. Using the information on package labels, compute the actual amount or percentage of fat, protein, and carbohydrates in various foods.

11

Career Applications and Career Project Ideas

Everyone is aware that calculators have cut the amount of time and effort involved in performing basic computations. For many professions, specific calculators are available that are able to perform the very specific functions used in that profession. However, one of the biggest advances—and probably the most potentially useful—is the advent of the programmable calculator. Now a calculator costing less than $100 can be purchased that can be preprogrammed to work very specific problems just by entering the data. The number of programs these calculators can perform is limited only by the time and talents of the programmer.

Business leaders can go into negotiations for companies, contracts, or new business ventures armed with calculators that have been preprogrammed to help them determine potential earning or selling costs or to figure installment payments. Prior to the advent of the programmable calculator, these same business leaders had to guess at the results of their negotiations or else recess for an extended period of time while the data were fed into a computer and then processed.

Sales personnel are able to consult their calculators to determine the best stock-option buys, to calculate installment payments, to determine interest rates and amounts, and to determine bond prices and yields.

Engineers were among the first to benefit from calculators. The development of the scientific calculator has made the slide rule, for years an indispensible tool of the engineer, obsolete.

Statisticians have calculators available that perform many of the operations involved in determining tests of significance. These calculators have the capability of determining sums, finding the sums of squares, storing information in their memories, and many other operations at just the touch of a button.

For people who often need to convert from the metric system to the customary system and back, specific calculators have been developed that make the conversions automatically. This is particularly useful for warehouse workers, purchasing agents, and people working with legal documents, such as land deeds.

Bankers, loan officers, and others interested in the loan business have calculators that replace all the books of tables that had to be used formerly. One of the great advantages of the calculator is that almost any chart or table derived from some formula, no matter how complex, can be printed into the circuitry of the calculator. All the operator has to do is enter the variables into the calculator and any answer that would be possible from the tables is instantly displayed on the calculator.

Pilots of both commercial and private airplanes have calculators that figure the time involved in the flight and help in determining the load balance, the fuel needs, and the course headings.

Druggists are using calculators that tell them the prescriptions, the amounts, and the correct dosages for patients who come in for refills. The memory and programmable features of calculators enable them to store and instantly recall any information previously fed into them.

Contractors are able to enter construction requirements, such as size, materials, time allotted, and other parameters into the preprogrammed calculator and come out with a bid price for the project.

Insurance agents and insurance adjustors find calculators essential in figuring rates, adjusting losses, and in doing the accounting associated with running such a business.

A great deal of the work a scientist does is related to mathematics. Formulas convert analyzed data into meaningful information. A calculator, with its memory capabilities, mathematical capability, and ability to be preprogrammed to solve almost any given formula, is an essential tool for any scientist.

This listing of professions and their uses of calculators is only a sampling of the current status. In the future the development of more and more capabilities for calculators will assure them an increasingly important role as a tool in all professions.

Career Project Ideas

This section provides numerous suggestions for class activities or projects. The teacher should select those most relevant to class interests and abilities as well as content goals. In each of these situations the teacher must provide the problem, materials, and data or have the student collect the materials.

1. Use a map to add the distances traveled on a summer trip.
2. Estimate the length of a river in a state and then compute a better estimate using map scale.
3. Subtract the amount of pollution caused by automobiles and/or industry from the total amount of air pollution.
4. Compute annual U.S. wheat income by multiplying production of crops by the current price.
5. Use the calculator to count the number of parks in your county or state.

6. Compute the average number of letters on a page and in several books.
7. Compute the average speed of a trip, given time and distance.
8. Figure the percentage of various crops produced in a particular area and represent this by a pie graph.
9. Compute the average rainfall for a month, temperature for a week, or wind velocity for 10 days.
10. Compute the percentage of the state population your local city contains.
11. Predict the percentage of students in your school that will go to college.
12. Compute the percentage of students who do not drink milk for lunch.
13. Set up a weather map using Celsius temperature and compute the average U.S. temperature for one day at noon by using the temperature of the largest city in each state.
14. Record the population of the cities of your state to thousands and add the total population of the cities. How does it compare to the exact sum of the cities' populations?
15. Set up a scale for drawing the solar system on a piece of notebook paper.
16. Compare costs of automobiles in 10-year intervals, from the time of their invention to today. Graph the results.
17. Compare the average speeds of automobiles in 10-year intervals from the time of their invention to today. Graph the results.
18. Take a poll of the favorite sports in class, and graph the results.
19. Compute the average number of major earthquakes per year, worldwide.
20. Determine the gear ratios in an average bicycle.
21. Chart the emigration pattern of people from various countries to the United States over the last 50 years.
22. Using the last four groups of census data, predict the population of your city or community in the next census.
23. Set up a class rummage sale and compute inventory, sales, operating costs, and so on.
24. Compute the size of your state in relation to the United States. Express your answer as a percentage.
25. Compare the land surface area in your state to the water surface area.
26. Compare rural and urban growth in your area for the last 50 years.
27. Graph farm production in your area over the last 100 years.
28. Compute the percentage of high school graduates in your city who attend the various state universities.

29. Compare the estimated cost of solar energy heating of homes to gas or electric heat.
30. Determine court fine collections and graph how the money is spent.
31. Draw a graph to show projected energy needs for the year 2000.
32. Graph the income of the top 10 corporations in the United States.
33. Make a pictograph to show the number of beds in several hospitals in the area.
34. Set up a balanced diet, with the number of calories needed for breakfast, lunch, and dinner.
35. Survey the class and draw a graph showing favorite foods.
36. Survey the class and draw a graph showing favorite subjects.
37. Compute the amount of water wasted in a year by a dripping faucet.
38. Compute relative humidity.
39. Experiment with the length of a pendulum and the number of completed oscillations. Compute the relationship.
40. Find the height of your school by comparing your shadow to the school's shadow.
41. Compute the amount of fat, protein, and so on in packaged food, using the information listed on the label.

12

Estimating with an Emphasis on Place Value

At times, calculator users take for granted that the computations performed are accurate. However, this is an assumption that should never be made. There are many opportunities to make errors on a calculator, and it is very important for the user to develop the habit of testing the reasonableness of the answers obtained.

With the calculator to give the exact answer, it is necessary for the user to be able to give only an approximate answer. The check of the estimate against the calculator's displayed answer gives an assurance of accuracy. Students should form the habit of estimating an answer for every problem worked. This estimate need not be recorded, but it can be used as a check on their work.

To gain facility in estimating, there are three concepts that must be understood, and skill must be developed in their use. First, both whole-number and decimal place-value concepts must be thoroughly understood. Second, skill in rounding numbers must be achieved. Third, the ability to work rapidly and accurately with multiples and powers of 10 must be developed.

If students can work all the problems in the check-out test for place value accurately, have them skip over to the check-out test on page 209 for rounding numbers.

Check-out test for place value

1. Write the number 50,305.025 in words.

2. Write the following as a numeral: Eight thousand four and three ten-thousandths.

3. Name the place value of the underscored digit.

a. $\underline{5}07{,}087$ **b.** $4.00\underline{2}6$ **c.** $22.9897\underline{7}$ **d.** $1.4\underline{0}9$

Answers:

1. Fifty thousand, three hundred five and twenty-five thousandths

2. 8,004.0003

3. **a.** hundred thousands
   b. thousandths
   c. hundred-thousandths
   d. hundredths

If you had problems with the check-out test for place value, work through the following section.

In our place-value system, each digit in a numeral has a place. Each place has a name. The places are grouped into *periods* of three places each. The number 537,112,280,665 has 12 places and 4 periods.

hundred billions	ten billions	billions	hundred millions	ten millions	millions	hundred thousands	ten thousands	thousands	hundreds	tens	ones
5	3	7,	1	1	2,	2	8	0,	6	6	5
billions			millions			thousands			ones		

We say that the 7 is in the billions place. The 8 is in the ten thousands place.

When we read large numbers we begin at the left and read one period at a time, pausing at each comma to name the period.

Try these:

Write the following numbers in words.

a. 75,802 **b.** 5,713,765 **c.** 549,380

Answers:

a. Seventy-five thousand, eight hundred two

b. Five million, seven hundred thirteen thousand, seven hundred sixty-five

c. Five hundred forty-nine thousand, three hundred eighty

When writing a numeral for a word number, every place must be accounted for by either a counting number or a zero acting as a place holder.

Try these:

Write the numerals for the following:

a. Forty thousand, twenty-six

b. Six million, thirty thousand, fifteen

c. Twenty-one million, three hundred four

Answers:

a. 40,026 **b.** 6,030,015 **c.** 21,000,304

We also write fractional numbers using place value. We use a decimal to separate the whole-number part and the fractional part. For example, 7.56 is another name for $7\frac{56}{100}$. A numeral like 7.56 is called a *decimal numeral.*

The place value of the numerals to the right of the decimal point begin with tenths.

ones		tenths	hundredths	thousandths	ten-thousandths	hundred-thousandths
1		9	3	6	7	4

To read a decimal like 1.93674, first read the digits, using the word *and* to name the decimal point, as a whole number. Then say the name of the last place. The numeral 1.93674 is read "one and ninety-three thousand, six hundred seventy-four hundred-thousandths."

Try these:

Write the word names for the following:

a. 0.4096 **b.** 5.065 **c.** 0.000021

Answers:

a. Four thousand ninety-six ten-thousandths

b. Five and sixty-five thousandths

c. Twenty-one millionths

To write the decimal numeral for a word name, write the number just as you would if it were a whole number, then count the number of decimal places needed. Add zeros on the right of the decimal point as needed for place holders for the decimal point.

Try these:

Write the decimal numerals for the following word names.

a. Twenty-six and forty-eight ten-thousandths

b. One thousand thirty-six and seven hundredths

c. Four ten-thousandths

Answers:

a. 26.0048 **b.** 1036.07 **c.** 0.0004

Check-out test for rounding

1. Round 762,342 to the nearest ten thousand.
2. Round 86,517 to the nearest thousand.
3. Round 19.355 to the nearest hundredth.
4. Round 0.94 to the nearest one.
5. Round the number 4765.6784

   a. to the nearest tenth; b. to the nearest thousand.

Answers:

1. 760,000 2. 87,000 3. 19.36 4. 1

5. a. 4,765.7 b. 5000

If you had problems with the check-out test for rounding, work through the following section.

The accompanying flow chart can be used to help make rounding any number an easy task.

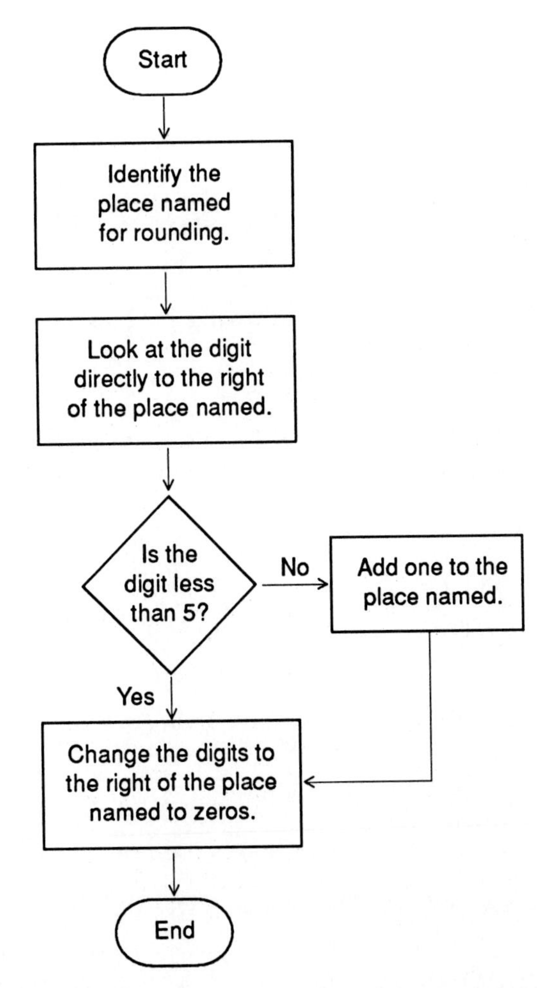

Follow the flow chart to round 10,271 to the nearest hundred.

Step 1. The 2 is in the place named for rounding (10,2̲71).

Step 2. The 7 is the digit directly to the right of the place named.

Step 3. The digit is not less than 5.

Step 4. Add 1 to the 2 of the place named.

Step 5. The 7 and the 1 are changed to zeros (10,300).

This same procedure can be used for rounding decimal numbers.

Try these:

Round the number 8,949.9873 to each specified place.

a. nearest thousand

b. nearest ten

c. nearest thousandth

d. nearest tenth

e. nearest hundredth

Answers:

a. 9,000 **b.** 8,950 **c.** 8,949.987 **d.** 8,950.0 **e.** 8,949.99

Check-out test for working with multiples and powers of 10

Write only the answers for the following problems.

1. 80,000,000 × 71,000

2. 20,000,000 ÷ 4000

3. 0.0001 ÷ 0.01

4. 0.001 × 0.0001

5. 50,000 + 1300

6. 700,000 - 9000

7. 24,000 × 600 ÷ 8,000

Answers:

1. 5,680,000,000,000 **2.** 5000 **3.** 0.01

4. 0.0000001 **5.** 51,300 **6.** 691,000 **7.** 1800

If you missed two or more problems, turn to Chapter 13 on scientific notation.

Once understanding and skill in using the concepts involved in place value, rounding, and working with multiples and powers of 10 have been achieved, becoming proficient in estimating is relatively easy.

Examples:

Use rounded numbers to find each sum or difference.

a. 180,217 + 204,348 = ?

200,000 + 200,000 = 400,000

180,217 + 204,348 is approximately equal to 400,000.

b. 2,201,022 - 750,021 = ?

2,200,000 - 800,000 = 1,400,000

2,201,022 - 750,021 is approximately equal to 1,400,000.

When deciding to which place value each number should be rounded, it usually is best to round all the numbers involved in addition and subtraction problems to the greatest place of the smallest number.

Try these:

Estimate the answers to the following problems and then check them with the Math Explorer calculator.

a. 5,767 + 32,892 + 38,599

b. 1,964,147 + 294,608

c. 1.503372 + 1,701.742 + 2,914,250

d. 40,600 - 38,964

e. 358.728 - 47.565

f. 1,023,110 - 976,875

g. 347,212 + 9,852,176

h. 17,698.537 + 750.021

i. 394,271 - 286,584

j. 358,728 - 10.987

k. 900,000 - 22.331

Answers:

a. 77,258

b. 2,258,755

c. 2,915,953.2

d. 1636

e. 311.163

f. 46,235

g. 10,199,388

h. 18,448.558

i. 107,687

j. 358,717.01

k. 899,977.67

When rounding numbers involved in multiplication and division problems, round each number involved to the largest nonzero place of each number.

Examples:

Use rounded numbers to find the products and quotients.

a. $680 \times 323 = ?$

$700 \times 300 = 210{,}000$.

680×323 is approximately equal to 210,000.

b. $2772 \div 37 = ?$

$3000 \div 40 = 75$

$2{,}772 \div 37$ is approximately equal to 75.

Try these:

Estimate the answers to the following problems and then check them with the Math Explorer calculator. (Hint: Count the number of decimal places in the number with the smallest number of decimal places. Press the [Fix] key, then enter that number in the Math Explorer.)

a. 4831×86

b. 672×545

c. $16{,}241 \times 234$

d. 914×4821

e. 6435×1786

f. 187×0.073

g. $124{,}986 \times 5.73$

h. 4.098×0.789

i. 358.976×1.487

j. $43{,}678 \times 0.095$

k. $274{,}963 \div 67$

l. $41{,}558 \div 742$

m. $161{,}774 \div 539$

n. $440{,}300 \div 88$

o. $139.2 \div 1.2$

p. $13.95 \div 0.05175$

q. $737.2 \div 0.07$

r. $0.3722 \div 9.36$

s. $760.5 \div 10.5$

t. $0.046 \div 0.0077$

Answers:

a. 415,466	**k.** 4103.9254
b. 366,240	**l.** 56.008086
c. 3,800,394	**m.** 300.13729
d. 4,406,394	**n.** 5003.4091
e. 11,492,910	**o.** 116
f. 13.651	**p.** 269.56522
g. 716,169.78	**q.** 10,531.429
h. 3.233322	**r.** 0.039765
i. 533.79731	**s.** 72.428571
j. 4149.41	**t.** 5.974026

13

Exponents and Scientific Notation

A calculator is a very useful tool for most computation, but sometimes it would be helpful if the display would show greater numbers. Most calculators today have an eight-digit display. The use of exponents and scientific notation can help the operator greatly increase the capability of a calculator if exact values are not essential. If students need to calculate exact values, have them work through student worksheets, *Expansion Team* and *Expansion Team II* in lesson 3 of Chapter 14. To begin, an understanding of exponential notation is essential.

An *exponent* is used to show the number of times that a number is used as a factor: $81 = 3 \times 3 \times 3 \times 3 = 3^4$. The factor (3 in this problem) is also called *the base.* 3^4 is read "three to the fourth power."

Try these:

Write the correct number of factors of 10 for each of the problems.

a. $10^4 = 10 \times 10 \times 10 \times 10$

b. 10^2

c. 10^7

d. 10^5

e. 10^1

f. 10^9

Write the following expressions using exponents.

g. $5 \times 5 \times 5 \times 5$

h. $2 \times 2 \times 2 \times 2 \times 2$

i. 3×3

j. $7 \times 7 \times 7 \times 7 \times 7 \times 7$

k. $10 \times 10 \times 10 \times 10 \times 10$

l. 9×9

Answers:

b. 10×10 **c.** $10 \times 10 \times 10 \times 10 \times 10 \times 10 \times 10$ **d.** $10 \times 10 \times 10 \times 10 \times 10$

e. 10 **f.** $10 \times 10 \times 10 \times 10 \times 10 \times 10 \times 10 \times 10 \times 10$

g. 5^4 **h.** 2^5 **i.** 3^2 **j.** 7^6 **k.** 10^5 **l.** 9^2

Supply the missing exponent:

m. $16 = 2^?$ **p.** $343 = 7^?$ **s.** $27 = 3^?$

n. $81 = 9^?$ **q.** $128 = 2^?$ **t.** $216 = 6^?$

o. $125 = 5^?$ **r.** $64 = 4^?$

Answers:

m. 4 **p.** 3 **s.** 3

n. 2 **q.** 7 **t.** 3

o. 3 **r.** 3

Write the next three numbers in each sequence.

u. Sequence of 1st powers is $1^1, 2^1, 3^1, 4^1,$ __, __, __

1, 2, 3, 4, __, __, __

v. Sequence of 2nd powers is $1^2, 2^2, 3^2, 4^2,$ __, __, __

1, 4, 9, 16, __, __, __

w. Sequence of 6th powers is $1^6, 2^6, 3^6, 4^6,$ __, __, __

1, 64, 729, 4096, __, __, __

x. Write the sequence of 3rd, 4th, and 5th powers.

y. Copy and complete this pattern:

$1^3 = 1^2$

$1^3 + 2^3 = 3^2$

$1^3 + 2^3 + 3^3 = \_^2$

$1^3 + 2^3 + 3^3 + \_^3 = \_^2$

Does the pattern in (y) hold for up to a base of 10? Can you give the rule for finding the base number that has a power of 2?

Answers:

u. $5^1, 6^1, 7^1$

5, 6, 7

v. $5^2, 6^2, 7^2$

$25, 36, 49$

w. $5^6, 6^6, 7^6$

$15{,}625, 46{,}656, 117{,}649$

x. $1^3, 2^3, 3^3, 4^3, 5^3, 6^3, 7^3$

$1^4, 2^4, 3^4, 4^4, 5^4, 6^4, 7^4$

$1^5, 2^5, 3^5, 4^5, 5^5, 6^5, 7^5$

y. 6^2

$4^3, 10^2$

The Math Explorer has the capability of performing the mathematics involved in evaluating most numbers in exponential form directly. Review Transparencies 23 and 24, *Powers of 10* and *Powers of Numbers* (Chapter 2) for help in working problems of this type on the Math Explorer calculator.

Try these:

Evaluate the following using the Math Explorer:

a. 5^5 **c.** $4^3 \times 4^4$ **e.** $7^7 \times 2^3$

b. $2^3 \times 5^3$ **d.** $(4 + 3)^3$ **f.** $(2 \times 4)^3$

Answers:

a. 3125 **c.** 16,384 **e.** 6,588,344

b. 1000 **d.** 343 **f.** 512

One of the most useful mathematics concepts for users of calculators is illustrated in the following examples.

a. $10{,}000 \times 100{,}000 = 10^4 \times 10^5 = 10^9 = 1{,}000{,}000{,}000$

b. $93{,}000{,}000 \times 3600 = 93 \times 1{,}000{,}000 \times 36 \times 100 =$

$93 \times 10^6 \times 36 \times 10^2 = 93 \times 36 \times 10^6 \times 10^2 =$

$3348 \times 10^8 = 334{,}800{,}000{,}000$

c. $100{,}000 \div 1000 = 10^5 \div 10^3 = 10^{5-9} = 10^2 = 100$

d. $688{,}800{,}000 \div 56{,}000 = 6888 \times 10^5 \div 56 \times 10^3 =$

$6888 \div 56 \times 10^{5-3} = 123 \times 10^2 = 12{,}300$

By expressing large or small numbers as products involving powers of 10 and working with the powers of 10 separately using the laws of exponents, the capability of the Math Explorer can be greatly increased.

Try these:

a. 224,000 × 623,000

b. 864,000 × 3650

c. 6,470,000 × 5280

d. 19,200,000 × 98,000

e. 4,056,000,000 ÷ 78,000

f. 4,250,000,000 ÷ 125,000

g. 513,300,000 ÷ 357,600

h. $(310,000)^2$

Answers:

a. 139,552,000,000

b. 3,153,600,000

c. 34,161,600,000

d. 1,881,600,000,000

e. 52,000

f. 34,000

g. 1,435.4027

h. 96,100,000,000

The division rule for exponents provides the following solutions when the denominator is greater than or equal to the numerator:

$10^3 \div 10^3 = 10^{3-3} = 10^0 = 1$
and
$100 \div 10,000 = 10^2 \div 10^4 = 10^{2-4} = 10^{-2} = 0.01$

Evaluate the following using the Math Explorer:

a. 5^5	**c.** 8^6	**e.** 3^8	**g.** 10^7
b. $5^2 \times 5^3$	**d.** $8^3 \times 8^3$	**f.** $3^5 \times 3^3$	**h.** $10^4 \times 10^3$

Answers:

a. 3125	**c.** 262,144	**e.** 6561	**g.** 10,000,000
b. 3125	**d.** 262,144	**f.** 6561	**h.** 10,000,000

The multiplication rule for numbers written in exponential form is illustrated by

$4^5 + 4^4 = 4^{5+4} = 4^9 = 262,144$ and $10^5 \times 10^3 = 10^{5+3} = 10^8 = 100,000,000$

Evaluate the following using the Math Explorer:

a. $6^5 \div 6^3$	**c.** $2^9 \div 2^6$	**e.** $8^6 \div 8^2$	**g.** $10^7 \div 10^2$
b. 6^2	**d.** 2^3	**f.** 8^4	**h.** 10^5

Answers:

a. 36	**c.** 8	**e.** 4096	**g.** 100,000
b. 36	**d.** 8	**f.** 4096	**h.** 100,000

The division rule for numbers written in exponential form is illustrated by

$4^5 \div 4^4 = 4^{5-4} = 4^1 = 4$ and $10^5 \div 10^3 = 10^{5-3} = 10^2 = 100$

Perform the following division problems. Express the answers as decimals or whole numbers.

a. $10^{11} \div 10^3$

b. $10^2 \div 10^6$

c. $10^7 \div 10^7$

d. $10^4 \times 10^2 \div 10^7$

e. $2^3 \times 2^4 \div 2^5$

f. $10^{-3} \times 10^4 \div 10^2 \times 10^{-1}$

g. $10^{-12} \div 10^{-5}$

Answers:

a. 100,000,000

b. 0.0001

c. 1

d. 0.1

e. 4

f. 1

g. 0.0000001

A decimal such as 0.0003 can be expressed as 3×10^{-4}. The exponent in the power of 10 is determined by the number of places the decimal point is moved in the problem. If the decimal is moved to the left, the exponent is positive: if it is moved to the right, the exponent is negative.

Try these:

Write each of the following as a number greater than or equal to 1 but less than 10 times a power of 10. For example, $0.0071 = 7.1 \times 10^{-3}$.

h. 0.002	**i.** 4,500	**j.** 0.0567	**k.** 23,000
l. 0.0001	**m.** 40.08	**n.** 120,000	**o.** 0.000024

Answers:

h. 2×10^{-3}	**i.** 4.5×10^3	**j.** 5.67×10^{-2}	**k.** 2.3×10^4
l. 1×10^{-4}	**m.** 4.008×10^1	**n.** 1.2×10^5	**o.** 2.4×10^{-5}

A number is said to be written in *scientific notation* when it is written as the product of a number greater than or equal to 1 and less than 10 and an integral power of 10. Scientific notation can be used to help perform computations with very large or very small numbers.

Try these:

Write the numbers in scientific notation and compute each sum or product. Express your answers in scientific notation, rounded to the nearest hundredth.

p. $0.00000000000000016019 \div 175{,}890{,}000{,}000$

q. $186{,}000 \times 31{,}536{,}000$

r. $0.00000000000667 \times 598{,}000{,}000{,}000{,}000{,}000{,}000{,}000 \times 80 \div (6400 \times 6400)$

s. $602{,}500{,}000{,}000{,}000{,}000{,}000{,}000 \times 0.0000000000000000001602$

Answers:

p. 9.11×10^{-27}	**q.** 5.87×10^{12}	**r.** 7.79×10^6	**s.** 9.65×10^4

14

Lesson Activities and Answers

Activities by Grade Level

K-3		4-6		7-9	
Lesson	Worksheet	Lesson	Worksheet	Lesson	Worksheet
2	Does It All Add Up? Added Value	2	Added Value Value Less What Number Gets You There? How Many in a Million? Coded Values	2	What Number Gets You There? How Many in a Million? Coded Values Changing Values
3	What Difference Does It Make? Varied Expressions Again?	3	Again? Add, Subtract, Multiply or Divide Pal of Whose? Gone Gosling Quick Sum Triangle Addition Pyramid Scheme Expansion Team Expansion Team II	3	Quick Sum Triangle Addition Expansion Team Expansion Team II
4	Add On Times, Times, Times	4	Times, Times, Times Which Pays? Perfectly Magic Constantly Yours	4	Which Pays? Perfectly Magic Constantly Yours On and On!
5	Find the Message	5	Math Order Well-Placed Parentheses The Calculator Rules Simply Put!	5	Well-Placed Parentheses The Calculator Rules Simply Put!
6	It's About Time!	6	It's About Time! Whole Number Divison True to Form Measuring Up and Down Rules Rule!	6	True to Form Measuring Up and Down Rules Rule!

K-3		4-6		7-9	
Lesson	Worksheet	Lesson	Worksheet	Lesson	Worksheet
7	What Is Your Group? Family Friends?	7	What Is Your Group? Family Friends? Subtraction and Division Order Workout! Mind Over Calculator	7	Workout! Mind Over Calculator
		8	Five Alone Endless Fractions?	8	Five Alone Endless Fractions? Formula Four Taxing Problems
		9	Step by Step	9	Step by Step Having a Party Separated In the Shade
		10	How Low Can You Go? Is It Simple? Going Up?	10	How Low Can You Go? Is It Simple? Going Up? Is It the GCF, the GED, or the SOP?
		11	Common Factors LCM	11	Common Factors LCM Euclid's Secret Eighth-Grade Discovery GCF
		12	Fraction Estimation Add and Subtract Fractions Multiply and Divide Fractions Rule of Thumb Rule of Thumb, Jr.	12	Rule of Thumb Rule of Thumb, Jr. Best Guess Good, Better, Best Tables of Stone?
		13	Checking Squares Ordered Drill	13	Checking Squares Ordered Drill Ups and Downs A Mixed Bag Improper Changes

K-3		4-6		7-9	
Lesson	Worksheet	Lesson	Worksheet	Lesson	Worksheet
		14	Another Mixed Bag The Trouble Maker Is It Really Magic?	14	Another Mixed Bag The Trouble Maker Is It Really Magic? Ancients of Noitcarf
		15	A Record Event Determination What Does That Point Prove? Redundancy Step Along	15	What Does That Point Prove? Redundancy Step Along Four Straight
		16	Decimal Fractions Completion Test Order Up! Converted Equivalent Values?	16	Decimal Fractions Completion Test Order Up! Converted Equivalent Values? Find the Fraction
		17	New Names 100 Per Learn the Symbol Percent Assigned Mixed Sequence	17	New Names 100 Per Learn the Symbol Percent Assigned Mixed Sequence
		18	Sun Country	18	Sun Country One to Another Raises and Such Commission er??
		19	Valid Relationships Circle Up! Do the Parts Equal the Whole?	19	Valid Relationships Circle Up! Do the Parts Equal the Whole? The Yummy Yogurt Marketing Meeting Square "Roots" Pizza Bargains Varied Dimensions
		20	Approximations Which Library is This? Who Was Pythagoras?	20	Approximations Which Library is This? Who Was Pythagoras? Power Match

K-3		4-6		7-9	
Lesson	Worksheet	Lesson	Worksheet	Lesson	Worksheet
		21	Upside Down	21	Upside Down
			Over or Under?		Over or Under?
			One Time!		One Time!
			Increased Dividends		Increased Dividends
		22	Notable Notations	22	Notable Notations
		23	Learn the Signs	23	Learn the Signs
			More Signs to Follow		More Signs to Follow
			Signs of the Times		Signs of the Times
			Signed On		Signed On
		24	Estimate the Area	24	Roaming Areas
			Roaming Areas		Long John Silver?
			Long John Silver?		Prison or Prism?
		25	Average Magic	25	Average Magic
			North to the Future?		North to the Future?
			Mean, Median, and Mode		Mean, Median, and Mode
					How Frequently Does it Occur?
		26	Millionaires!	26	Millionaires!
			Building Fences		Building Fences
			Body Works!		Body Works!
			Math Explorer Overflow		Math Explorer Overflow
			Only Two Keys Fit		Only Two Keys Fit
					Did You Get That?
					Check Your Answers!
					Calendar Magic
					Temperature Conversion

Lesson 1: Introduction to the Math Explorer

Overview

The student learns the overall layout of the calculator and how to use the clearing keys.

Transparencies

Transparencies 3 and 4, *Keyboard* and *Clearing Keys,* Chapter 2

Keys Introduced

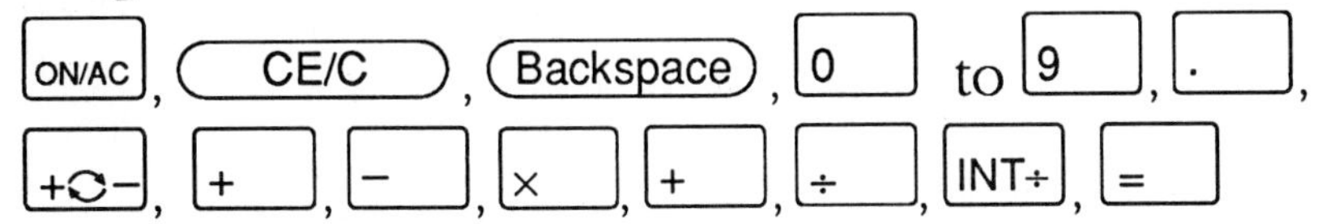

Student Worksheets

None

Teaching Steps

1. Use Transparency 3, *Keyboard*, to show students where these keys are located:
   - Digit keys: [0] to [9], [.], [+⇄−]
   - Operations keys: [+], [−], [×], [÷], [INT÷], [=]
   - [ON/AC], (CE/C), (Backspace)
2. Use Transparency 4, *ON/AC, Backspace,* to show students how to clear an entry by pressing the (CE/C) key.
3. Have the students enter a series of numbers like 124678 and practice using the (Backspace) key to erase one or more of the numbers.

Five-Minute Fillers

Make all the words you can from the letters in CALCULATOR.

List all the words that can be formed on your calculator from the letters L, G, S, H, E, O, I, and B. These are the letters formed by the numbers when you read the display upside down.

Lesson 2: Place Value

Overview

Students learn how to use the calculator to determine place value and to identify the digit occupying a given place.

Transparencies

Transparencies 16 and 26, *Fixing the Decimal Point* and *Parentheses*, Chapter 2

Keys Introduced

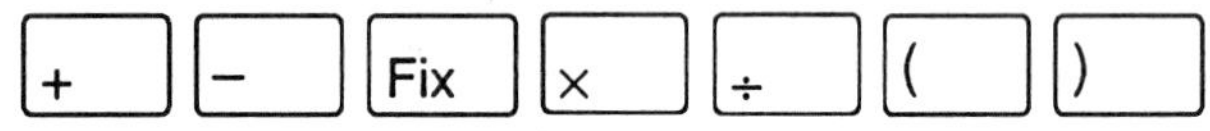

Student Worksheets

Does It All Add Up?
Added Value
Value Less
What Number Gets You There?
How Many in a Million?
Coded Values
Changing Values

Teaching Steps

1. Use Transparencies 16 and 26, *Fixing the Decimal Point* and *Parentheses* to demonstrate how to set the number of desired points and to group numbers in desired groupings.
2. Have students work the *Try these* on the transparencies.
3. Students will have some understanding of place value, but work closely with them through these exercises to be sure they understand the concepts.

Five-Minute Fillers

1. Find the sums:

987,654,321	123,456,789
87,654,321	12,345,678
7,654,321	1,234,567
654,321	123,456
54,321	12,345
4,321	1,234
321	123
21	12
+ 1	+ 1

2. Write all the three-digit numbers you can using the digits 3, 6, and 9 once in each number (there are 6 numbers). Subtract each of these numbers from 1332. What do you notice?

Does It All Add Up?

Directions

1. Draw a circle around the numeral that shows the number of counters in the cans in each row.

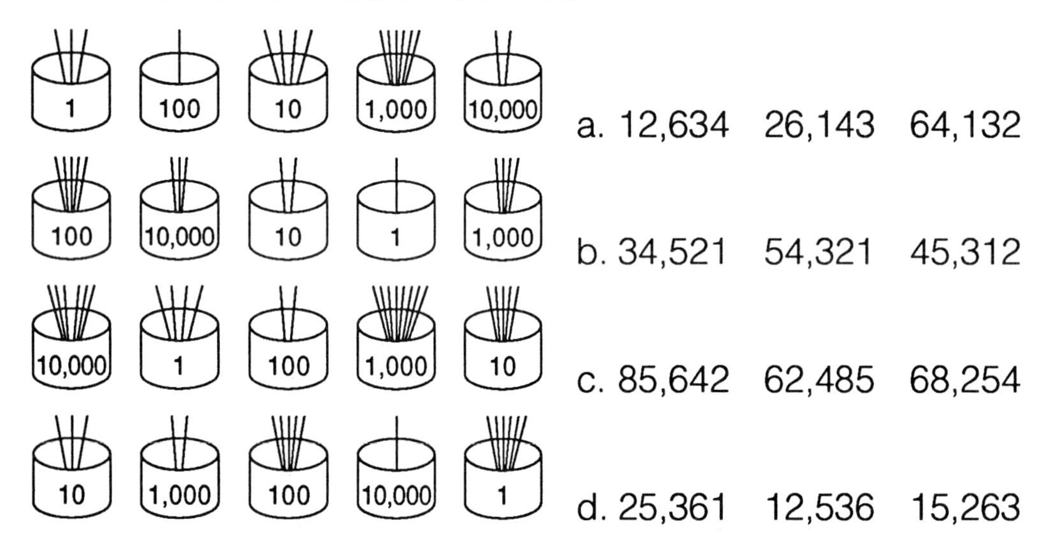

2. Add all the circled answers on the Math Explorer. Do you get an answer of 141,454? ________

Added Value

Directions

1. This game is played by two teams of eight students each.
2. The teacher writes an eight-digit numeral on the chalkboard (each digit is used only once), such as

   32,718,469

3. The two teams are in rows; the student at one end of each team has a Math Explorer. The students must play this game in silence.
4. The teacher calls out a number in the eight-digit numeral and the student "adds" it on the Math Explorer, with the correct place value. For example, 8 is in the thousands place, so the student would record 8000 in the Math Explorer.
5. The student passes the Math Explorer to the next team member, and the teacher calls out another number. The student adds the number with the correct place value, and passes it on. For example, if the teacher called out 7, the student would add 700,000.
6. When the eighth number has been added by the eighth team member, the numeral should be the same as the numeral on the board.
7. The team that has the same numeral as the one on the board gets 1 point. If both teams have the correct numeral, they both get a point.
8. After five times, the team with the highest score wins the game.

Value Less

Directions

1. Enter each number in the Math Explorer.
2. Enter an operation and a number to remove the place value number indicated.

Number	Place value	Result
a. 137	3 tens	107
b. 1,475	7 tens	______
c. 3,651	3 thousands	______
d. 2301	2 ones	______
e. 436	3 tens and 6 ones	______
f. 4.37	3 tenths	______
g. 1,375	1 thousand and 7 tens	______
h. 4,675,432	67 ten thousands	______
i. 7,967,732	the sevens	______
j. 376,543	the threes	______
k. 14,441	the fours	______
l. 14,441	the ones	______
m. 8,635,765	the sixes	______

What Number Gets You There?

Directions

1. Enter the number 1,000,000 into the display of the Math Explorer, then press the (CE/C) key.

2. Enter the number 1000 into the display of the Math Explorer.

3. Guess what operation key, [+], [−], [×] or [÷], and what number you would now have to enter to get the number 1,000,000 in the display again.

   ____

4. Try your guess. Keep guessing until you get the number 1,000,000 in the display.

5. Can you find another operation and another number you can use on the 1000 in the display to have an answer of 1,000,000 appear again?

6. Decide what operation and what number you would have to enter to get 1,000,000 if you first enter 2000 or 5000 or 10,000 or 750 in the display.

   ____

How Many in a Million?

Directions

1. Enter the number 1,000,000 into the display of the Math Explorer.

2. Guess what operation key you would have to press and what number key you would have to press to have the number 1000 appear in the display of the Math Explorer.

   ____

3. Keep trying until you are successful in getting the number 1,000 to appear in the display.

4. Can you find another operation and another number that will give you the same results?

5. Decide what operation and what number you would have to enter to change the number 1,000,000 in the display to 2000 or 5000 or 10,000, or 750.

   ____

Coded Values

Directions

1. Work each of the problems.
2. Round each problem to the place value indicated using the [Fix] key.
3. Turn the Math Explorer upside down to read the word given.
4. Write the word given by the Math Explorer.

   a. (1543 + 2324) × 0.0002 (ten thousandths) __________

   b. 48,999 ÷ 69 (tens) __________

   c. 0.05025 × 16 (thousandths) __________

   d. 69.2225 × 8.00001 (hundredths) __________

   e. 0.58345 × 6.00003 (ten thousandths) __________

   f. 2.3 × 1.9598 (ten thousandths) __________

   g. 268.40937 ÷ 50.6001 (ten thousandths) __________

Changing Values

Directions

1. Enter a number such as 578.907 in the Math Explorer.
2. Use one operation and one number to change that number to 578.937.
3. Practice by making the following changes:
   a. 36.519 to 36.513 ________
   b. 2.03125 to 2.03128 ________
   c. 31,188 to 32,188 ________
   d. 0.1953016 to 0.2953016 ________
   e. 0.14287 to 0.14687 ________
   f. 0.0555 to 0.0565 ________

Lesson 3: Add, Subtract, Multiply, or Divide Numbers

Overview

Students learn how to use the calculator to add, subtract, multiply, and divide whole numbers. Depending on the grade level you are teaching, you can also show students how to use the [=] key to repeat an operation.

Transparencies

Transparencies 5 and 9, *Math Operations* and *Repeat Operations*, Chapter 2

Keys Introduced

None

Student Worksheets

What Difference Does It Make?
Varied Expressions
Again?
Add, Subtract, Multiply, or Divide
Pal of Whose?
Quick Sum
Triangle Addition
Pyramid Scheme
Expansion Team
Gone Gosling
Expansion Team II

Teaching Steps

1. Use Transparency 5, *Math Operations*, to demonstrate how to use the operation keys.
2. Have students work the *Try these* on the transparency.
3. If you are working with decimals and you have already taught how to fix the decimal place, have students do problems with different places set for the decimal.

4. If appropriate, use Transparency 9, *Repeat Operations*, to show students how to use the calculator to repeat a calculation. The built-in constant works with these keys: [+] [−], [×], [÷] and [INT÷].

5. Show students how to clear underflow or overflow errors:

   - If calculations result in answers that are too large for the calculator, **Error O** appears in the display to indicate overflow.

   - If calculations result in answers that are too small for the calculator, **Error U** appears in the display to indicate underflow.

   Examples:
   Multiply 250,000 × 10,000
   Multiply 0.00005 × 0.00003

   - To clear *any* error condition that appears in the display, press [CE/C] [CE/C] or [ON/AC] and reenter the entire calculation.

6. Complete the examples on the transparency.

Five-Minute Fillers

1. Find three numbers that divide 2431 evenly.

2. Choose any seven consecutive numbers. Find their sum. Multiply the fourth, or middle number, by 7. Compare the results.

3. Find the sum of any eleven consecutive numbers by multiplying.

4. Find as many ways as you can to combine the digits 1 through 9, in order, to equal 100. (Example: 12 + 3 + 4 + 5 - 6 - 7 + 89 = 100)

5. Find as many ways as you can to combine the digits 9 through 1, in order, to equal 100. (Example: 9 - 8 + 76 + 54 - 32 + 1 = 100)

6. Choose a two-digit number and write that number three times to form a six-digit number. (Example: 29 — 292,929) Divide your number by 37, then 13, then 7, then 3. Will the result always be the same?

What Difference Does It Make?

Directions

Use the Math Explorer to solve the cross-number puzzle.

Be sure you understand the self-checking aspect of the puzzle.

Across

1. 100 - 46
3. 103 - 9
5. 53 - 18
7. 96 - 22
8. 80 - 19
9. 54 - 27
10. 83 - 35
12. 410 - 301
13. 75 - 58

Down

1. 75 - 18
2. 861 - 417
3. 125 - 29
4. 70 - 29
5. 500 - 180
6. 862 - 283
11. 90 - 9
12. 30 - 11

Varied Expressions

Directions

Experiment with the Math Explorer to answer the following.

1. The number 10 is also

_______ + _______

_______ − _______

_______ × _______

_______ ÷ _______

2. The number 83 is also

_______ + _______

_______ − _______

_______ × _______

_______ ÷ _______

3. The number 67 is also

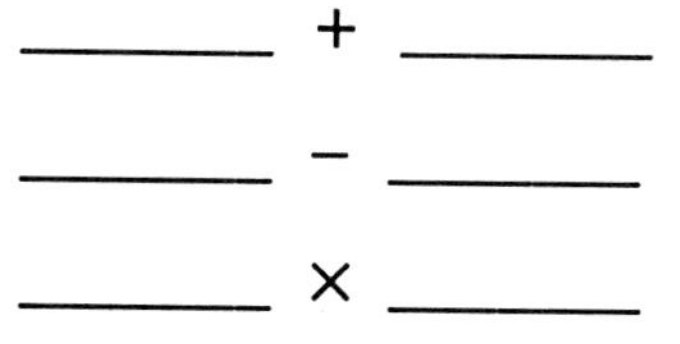

_______ + _______

_______ − _______

_______ × _______

_______ ÷ _______

4. The number 132 is also

_______ + _______

_______ − _______

_______ × _______

_______ ÷ _______

5. The number 77 is also

_______ + _______

_______ − _______

_______ × _______

_______ ÷ _______

6. The number 111 is also

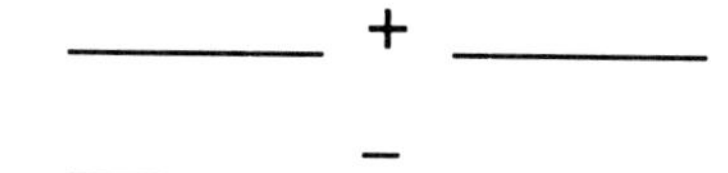

_______ + _______

_______ − _______

_______ × _______

_______ ÷ _______

Again?

Directions

1. Select any three-digit number in which the first digit is larger than the last.
2. Reverse the digits and subtract the second number from the first number.
3. Reverse your answer from step 2 and add that number to the answer from step 2.
4. Do this at least four times.
5. Compare your answers.
6. Can you find a case in which the answer is different when you follow steps 1-3?

Example:

```
 782
-287
----
 495
+594
----
 ???
```

Add, Subtract, Multiply, or Divide

Directions

Add, subtract, multiply or divide.

1. 12 + 15 = ______
2. 453 + 654 = ______
3. 5 + 22 = ______
4. 138 + 2513 = ______
5. 71 + 35 = ______
6. 67 + 908 = ______
7. 123 + 231 = ______
8. 141 + 14 = ______
9. 15 - 12 = ______
10. 4567 - 3467 = ______
11. 121 - 34= ______
12. 3424 - 256 = ______
13. 71 - 35 = ______
14. 35 - 21 = ______
15. 831 - 113 = ______
16. 173 - 167 = ______
17. 15 × 12 = ______
18. 14 × 425 = ______
19. 121 × 34 = ______
20. 124 × 345 = ______
21. 71 × 35 = ______
22. 5631 × 4256 = ______
23. 290 × 113 = ______
24. 3711 × 7211 = ______
25. 12 ÷ 15 = ______
26. 45 ÷ 6 = ______
27. 34 ÷ 13 = ______
28. 32 ÷ 17 = ______
29. 231 ÷ 211 = ______
30. 1358 ÷ 231 = ______
31. 26 ÷ 4 = ______
32. 4315 ÷ 973 = ______

Pal of Whose?

Directions

1. Enter any number of 2 or more digits into the Math Explorer.
2. Reverse the digits and add the new number to the one already in the display.
3. Repeat step 2 until you get a number that reads exactly the same from right to left as from left to right.

   A number that reads the same both ways is called a palindrome.

   Example:

   $$\begin{array}{r} 568 \\ +\ 865 \\ \hline 1433 \\ +\ 3341 \\ \hline 4774 \end{array}$$

4. Find the palindromes of each of these numbers:

   a. 342 __________

   b. 7586 __________

   c. 9644 __________

   d. 3687 __________

   e. 1908 __________

5. Can you find a two-digit number that fills the screen before it becomes a palindrome?

   ____

Gone Gosling

Directions

Once you have worked the problem, turn the Math Explorer upside down to answer the question.

1. Sue had a friend by the name of _____________.
   54,026 ÷ 7
2. What did Sue say when she saw him? _____________
   3867 ÷ 5000
3. What did Bill sit on that made him jump? _____________
   6422 ÷ 19
4. What did Bill want to be when he grew up? _____________
   1206 ÷ 1500
5. Sue said that if that was what he wanted to be

   she thought he was a real what? _____________

   87,515 ÷ 25,000

Quick Sum

Directions

1. Select any two numbers and record them.

 ____

2. Get the third addend by using the Math Explorer to find the sum of the first two addends.

 ____

3. Get the fourth addend by using the Math Explorer to find the sum of the second and third addends.

 ____

4. Continue getting each new addend by using the Math Explorer to add the previous 2 addends until there are 10 addends in the problem.

 ________ ________ ________ ________ ________ ________

5. The sum of all 10 addends can be obtained by using the Math Explorer to multiply the seventh addend by 11. Check your answer by adding the column on the Math Explorer.

Example:

564	[first number]
852	[second number}
1,416	[sum of first two numbers]
2,268	[sum of two previous numbers]
3,684	
5,952	
9,636	} 9636 × 11 = 105,996
15,588	
25,224	
+40,812	
105,996	

6. Try this with several other pairs of numbers.

Triangle Addition

Directions

1. Use the Math Explorer to find the sum of each row in the triangles below.
2. Can you find a pattern? Use that pattern to find the sum of the next four rows before you write the rows.
3. Write the next four rows.
4. Find the sums of the diagonals each way. Are there any patterns?

1 _____

3 5 _____

7 9 11 _____

13 15 17 19 _____

21 23 25 27 29 _____

31 33 35 37 39 41 _____

2 _____

4 6 _____

8 10 12 _____

14 16 18 20 _____

22 24 26 28 30 _____

32 34 36 38 40 42 _____

Pyramid Scheme

Directions

Use the Math Explorer to find the patterns. If a number becomes too large for the Math Explorer, try to use the pattern to find the remaining answers.

$1 \times 1 =$ ______

$11 \times 11 =$ ______

$111 \times 111 =$ ______

$1111 \times 1111 =$ ______

$11111 \times 11111 =$ ______

$111111 \times 111111 =$ ______

$1111111 \times 1111111 =$ ______

$11111111 \times 11111111 =$ ______

$1 \times 9 + 2 =$ ______

$12 \times 9 + 3 =$ ______

$123 \times 9 + 4 =$ ______

$1234 \times 9 + 5 =$ ______

$12345 \times 9 + 6 =$ ______

$123456 \times 9 + 7 =$ ______

$1234567 \times 9 + 8 =$ ______

$12345678 \times 9 + 9 =$ ______

$123456789 \times 9 + 10 =$ ______

Expansion Team

Directions

1. When a problem is too big for the Math Explorer display, such as:

$$\begin{array}{r} 1{,}364{,}457{,}175 \\ \times 43 \\ \hline \end{array}$$

Use the distributive law to find the answer: $x(y + z) = xy + xz$.

$$43 \times (1{,}364{,}457{,}175) \rightarrow \begin{array}{r} 1{,}364{,}000{,}000 \\ \times 43 \\ \hline \end{array} + \begin{array}{r} 457{,}175 \\ \times 43 \\ \hline \end{array}$$

$$\begin{array}{r} 1{,}364 \times 10^6 \\ \times 43 \\ \hline 58{,}652 \times 10^6 \end{array} + \begin{array}{r} 457{,}175 \\ \times 43 \\ \hline 19{,}658{,}525 \end{array}$$

$$\begin{array}{r} 58{,}652{,}000{,}000 \\ +\ 19{,}658{,}525 \\ \hline 58{,}671{,}658{,}525 \end{array}$$

2. Solve the following the same way:

a. 3,641,589,211 × 63 ____________________

b. 4,987,632,118 × 27 ____________________

c. 3,654,136 × 127 ____________________

d. 4,699,192 × 715 ____________________

e. 5,999,871 × 987 ____________________

Expansion Team II

Directions

1. Follow these steps if the number is too large for the Math Explorer.

   Example: 365,413,996 ÷ 6

   Step 1: Make the problem smaller (enter the first 5 digits into the Math Explorer).

   36,541

   Step 2: Divide using the [INT÷] key so that your remainder is expressed as a whole number. Record the whole-number quotient.

   36,541 [INT÷] 6 = 6090 R1

   Step 3: Place your remainder to the left of the remaining digits of the original dividend and divide by the divisor again.

   13,996 [INT÷] 6 = 2332 R4

   Step 4: Combine the whole number answers from step 2 and step 3, then add the remainder from step 3.

   60,902,332 R4

2. Use the preceding steps to solve the following division problems on the Math Explorer.

   a. 418,376,514 ÷ 8 ____________________

   b. 986,143,552 ÷ 7 ____________________

   c. 178,694,237 ÷ 56 ____________________

   d. 75,241,157 ÷ 83 ____________________

   e. 872,154,771 ÷ 137 ____________________

Lesson 4: Using the Constant Feature

Overview

Students learn how to use the [Cons] key to set up the counter. Students also learn how to use the [Cons] key to build an addition table.

Transparencies

Transparency 10, *Constant Counter*, Chapter 2

Keys Introduced

[Cons]

Student Worksheets

Add On
Times, Times, Times
Which Pays?
Perfectly Magic
Constantly Yours
On and On!

Teaching Steps

1. Use Transparency 10, *Constant Counter*, to show students how to use the [Cons] key to set up the counter:
   - Enter the operation you want: [+], [−], [×], [÷] or [INT÷].
   - Select the number you want to add, subtract, multiply by, divide by, or integer divide by.You can enter a whole number, decimal, or fraction for that number.
   - Set up the constant operator by pressing [Cons].
   - Enter the number on which you want to operate.
   - Continue to press the [Cons] key to operate on the selected number.
2. Show students that **Error C** appears in the display to indicate a [Cons] key error.

Press	**Display**
2 [+] [Cons]	Error C

3. Point out that the CONS and + indicators appear in the display when [Cons] is pressed. Also point out that the first number in the display is the number being added to by the constant and the second number is the answer.
4. Have students work several problems to get familiar with the process.

Five-Minute Filler

Follow these steps to a shortcut. Press only the keys shown.

9 [×] 12 [=] *108*

7 = ________

11 = ________

20 = ________

4 = ________

100 = ________

Follow the same steps to multiply

54 by 2 ________

by 10 ________

by 8 ________

by 5 ________

Record your conclusions!

Add On

Directions

Complete the following addition sequences by following the directions for each problem.

1. Press the [+] key, press the [1] [0] keys, and press the [Cons] key. Now enter 10 into the display of the Math Explorer. Then press the [Cons] key once for each space that needs an answer.

   10, ____, ____, ____, ____, ____, ____, ____, ____, 100

2. Press the [+] key, press the [5] key, and press the [Cons] key. Now enter 0 into the display of the Math Explorer. Then press the [Cons] key once for each space that needs an answer.

   0, 5, ____, ____, ____, ____, ____, ____, ____, ____, ____, ____, 60

3. Use the procedure described in steps 1 and 2 to complete the following sequences.

   a. 2, then 0, ____, ____, ____, ____, ____, ____, ____, ____, ____, ____, 22

   b. 3, then 0, ____, ____, ____, ____, ____, ____, ____, ____, ____, ____, ____, 36

   c. 4, then 0, ____, ____, ____, ____, ____, ____, ____, ____, ____, ____, 44

   d. 25, then 0, ____, ____, ____, ____, ____, ____, ____, ____, ____,

Times, Times, Times

Directions

Complete the following multiplication sequences by following the directions for each problem.

1. Press the [×] key, press the [5] key, and press the [Cons] key. Now enter 5 into the display of the Math Explorer. Press the [Cons] key for each space that needs an answer.

   5, ______, ______, ______, ______, ______, ______, 390,625

2. Press the [×] key, press the [3] key, and press the [Cons] key. Now press the [Cons] key for each space that needs an answer.

   3, ______, ______, ______, ______, ______, ______, 6,561

3. Use the procedure described in steps 1 and 2 to complete the following sequences.

   4, ________, ________, ________, ________, 4096

   7, ________, ________, ________, ________, 117,649

Which Pays?

Directions

Rubin's employer offered to pay him $300 for 15 days' work or to pay him 1 penny the first day, 2 pennies the second day, 4 pennies the third day, 8 pennies the fourth day, and so on, up to 15 days. Which offer should Rubin take? How much would he make taking the second offer?

Day 1: 1 cents

Day 2: 2 cents

Day 3: 4 cents

Day 4: 8 cents

Day 5: ________

Day 6: ________

Day 7: ________

Day 8: ________

Day 9: ________

Day 10: ________

Day 11: ________

Day 12: ________

Day 13: ________

Day 14: ________

Day 15: ________

What is the biggest wage Rubin could earn? ________

Perfectly Magic

Directions

1. All the magic squares shown are multiples of the completed square.
2. A square is magic if the sum when you add vertically, horizontally, or diagonally is the same number.
3. Complete the magic squares using the Math Explorer by following these steps to find the missing numbers:

7 [×] 7 = 49

___ ☐ 6 = ____

___ ☐ 9 = ____

6	7	2
1	5	9
8	3	4

	49	

	44.1	
50.4		25.2

78		
		117
	39	

Constantly Yours

Directions

Use the [×] key on the Math Explorer to help find the patterns.

Enter 2 [×] 999,999 = ____________,

3 = ____________,

4 = ____________,...

999,999 × 2 = ____________	99 × 12 = ____________
999,999 × 3 = ____________	99 × 23 = ____________
999,999 × 4 = ____________	99 × 34 = ____________
999,999 × 5 = ____________	99 × 45 = ____________
999,999 × 6 = ____________	99 × 56 = ____________
999,999 × 7 = ____________	99 × 67 = ____________
999,999 × 8 = ____________	99 × 78 = ____________
999,999 × 9 = ____________	99 × 89 = ____________

On and On!

Directions

1. Find the operation and the constant needed to complete each sequence.
2. Use the constant key and the appropriate operation key to complete each sequence.

   a. 2, 9, 16, 23, ______, ______, ______, ______

   b. 7, 3, -1, -5, ______, ______, ______, ______

   c. 1.7, 2.5, 3.3, 4.1, ______, ______, ______, ______

   d. 2, 8, 32, 128, ______, ______, ______, ______

   e. 27, 9, 3, 1, ______, ______, ______, ______

   f. 2, 6, 18, 54, ______, ______, ______, ______

   g. 16, 8, 4, 2, ______, ______, ______, ______

   h. 1, -2, 4, -8, ______, ______, ______, ______

Lesson 5: Order of Operations

Overview

Students learn how to do calculations with two or more operations. They also learn the priority in which calculations are accomplished by the calculator and the difference that the order of entry can make in the answer.

Transparencies

Transparency 8, *Order of Operations*, Chapter 2

Keys Introduced

None

Student Worksheets

Find the Message
Math Order
Well-Placed Parentheses
The Calculator Rules
Simply Put!

Teaching Steps

1. Use Transparency 7, *Order of Operations*, to demonstrate the order in which calculations are completed.

2. Explain how the AOS (Algebraic Operating System) works:

Priority	Keys	Comment
1	[(] [)]	Any operations in parentheses take priority over operations outside the parentheses.
2	[x^2] [$\sqrt{\ }$] [10^n] [1/x] [%]	Immediate functions — keys that operate on a single variable — are the next priority.
3	[y^x]	

4	[×] [÷] [INT÷]	Algebraic hierarchy says multiply and divide before adding and subtracting.
5	[+] [−]	
6	[=]	Completes all pending operations.

3. Explain that the parentheses keys are a way around the algebraic hierarchy. Although parentheses are technically not part of the algebraic hierarchy, students need to understand that any operations within parentheses take priority over operations outside of the parentheses.
4. Review the priority of operations that the calculator uses and how that priority can make a difference in the answer.
5. Discuss with students the order in which the operations in the *Try these* examples are completed instead of having students work them on the calculator.
6. Have students complete Transparency 7.

Five-Minute Filler

What two important dates are given by the problem 48 × 64,318 ÷ 2 × 29 ÷ 3?

Find the Message

Directions

Work the problems below. Then use the decoder at the bottom of this page to match your answers to the letters to discover the hidden message.

Problem		Answer	Letter
a. 16 × 3 ÷ 2	=	______	______
b. 15 - 5 × 3	=	______	______
c. 1 + 3 × 7	=	______	______
d. 5 - 22 ÷ 11	=	______	______
e. 123 ÷ 3 × 2	=	______	______
f. 3 × 4 + 6	=	______	______
g. 3 - 4 + 6	=	______	______
h. 3 + 4 × 6	=	______	______
i. 12 ÷ 2 - 2	=	______	______

Decoder

This answer:	Is this letter:
0	A
3	H
4	N
5	F
18	S
22	T
24	M
27	U
82	I

Math Order

Directions

Solve the problem and then verify with the Math Explorer calculator.

	Your Answer	Calculator answer
$2 + 4 \times 5 =$	______	______
$12 + 6 \div 2 =$	______	______
$8 \times 4 - 5 =$	______	______
$12 \div 6 + 2 =$	______	______

What is the order of operations?

$9 \div 3 + 6 \div 2 =$	______	______
$9 \times 3 + 6 \times 2 =$	______	______
$9 - 3 + 6 \div 2 =$	______	______
$9 - 3 + 6 - 2 =$	______	______
$9 - 3 \times 6 \times 2 =$	______	______
$9 + 3 - 6 \div 2 =$	______	______

What is the order of operations?

$16 \div 4 \div 2 \times 3 =$	______	______
$6 \times 4 \times 2 \div 2 =$	______	______
$16 \div 4 - 2 \times 3 =$	______	______
$16 \div (4 - 2) \times 3 =$	______	______

Conclusion: The order of operations is ______________________

____

Well-Placed Parentheses

Directions

1. Insert parentheses into each expression to make the number sentence true. Verify your placement using the Math Explorer and the (,) keys.

   a. $4 \times 3 + 3 \times 4 = 24$

   b. $8 \div 4 + 4 \div 2 = 4$

   c. $4 \times 3 + 3 \times 4 = 96$

   d. $8 \div 4 + 4 \div 2 = 3$

   e. $4 \times 3 + 3 \times 4 = 60$

   f. $8 \times 8 \div 8 \times 8 - 8 + 8 = 64$

   g. $6 + 5 - 4 - 3 = 10$

   h. $8 \times 8 \div 8 \times 8 - 8 + 8 = 8$

   i. $6 + 5 - 4 - 3 = 4$

   j. $42 \div 6 - 6 + 13 \times 13 = 170$

2. Now use exactly four 4s to name the numbers from 1 through 12. [**Example:** $(4 \div 4) + (4 - 4) = 1$]

The Calculator Rules

Directions

1. Solve these problems using the Math Explorer. Compare your answers in each pair of problems. What do the parentheses tell the Math Explorer to do?

   a. $56 \div 4 \div 2 = 7$

   b. $56 \div (4 \div 2) = 28$

   c. $5 \times 4 + 4 =$ ______

   d. $5 \times (4 + 4) =$ ______

   e. $80 - 9 - 6 =$ ______

   f. $80 - (9 - 6) =$ ______

   g. $(2 \times 5 + 6) \times 4 =$ ______

   h. $(2 \times 5) + (6 \times 4) =$ ______

   i. $7 \times 4 - 1 =$ ______

   j. $7 \times (4 - 1) =$ ______

2. Using the Math Explorer, fill in the parentheses to make the sentence true.

   a. $84 \div 21 \div 3 = 12$

   b. $4569 + 596 \div 4 - 38 \times 16 = 4110$

   c. $859{,}635 - 651 \times 937 - 4962 \div 827 = 249{,}642$

   d. $9104 \times 3675 - 4762 \times 2352 = 22{,}256{,}976$

   e. $451{,}364 + 460{,}924 \div 3521 + 3411 = 131.60531$

Simply Put!

Directions

1. Simplify (without the aid of the Math Explorer) each of the following expressions. If there is no operation symbol between a number and a parenthesis symbol, multiply.

   a. $3 \times 4 + 6 \times 2$
   b. $3 + 3 \times 5 + 3 \times 6$
   c. $5 + 1 \times 6 - 9$
   d. $15 - 4 \times 6 + 10$
   e. $5(13 - 3 \times 4 + 2)$
   f. $12 + 8(3 + 4 \times 7)$
   g. $3[(12 - 3) \times 4 + 2]$
   h. $(9 + 4) \times (3 + 3 \times 6 - 5)$
   i. $4[7 + (5 + 4)]$
   j. $(11 + 3) \times 4(15 - 3 \times 4)$

2. Check your work with the Math Explorer.

3. Answer the following questions:

   a. Should you work within parentheses before you multiply?

      ______

   b. Should you add before you multiply? ______

   c. Should you operate left to right without regard to the sign?

      ____

   d. Which operation should you do first? Circle it.
      1. Addition or subtraction
      2. Multiplication or division
      3. Parentheses

   e. Which of the above operations should you do next?

      ______

   f. Which of the above operations should you do last?

      ______

Lesson 6:
Division with Remainders

Overview

Students learn to use the [INT÷] key to perform division with remainders. Then they can practice division problems on the *Whole Number Division* worksheet. If you are working with calculations involving units of time, there is an additional transparency and student worksheet to teach time calculations.

Transparencies

Transparencies 6 and 7, *Dividing Whole Numbers* and *Equivalent Units of Time*, Chapter 2

Keys Introduced

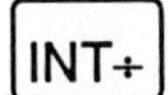

Student Worksheets

It's About Time!
Whole-Number Division
True to Form
Measuring Up and Down
Rules Rule!

Teaching Steps

1. Use Transparency 6, *Dividing Whole Numbers*, to demonstrate how to do whole number division with the [INT÷] key.
2. Explain the format of the answer:
   - The quotient is displayed first and is marked by **Q** and a bracket.
   - The remainder is displayed second and is marked by **R** and a bracket.
3. Remind students that all numbers used with the [INT÷] key must be positive whole numbers, or **Error I** appears in the display to indicate an integer error.

Press	Display
6 [INT÷] 3.5 [=]	Error I

4. Point out that the **I** indicator appears in the display when [INT÷] is pressed.
5. Tell students that if they use the result to perform another calculation, only the quotient is used; the remainder is dropped.

6. Have students complete Transparency 6. Remind them to clear the calculator by pressing CE/C or ON/AC before entering any calculations.
7. If you are working with calculations involving units of time, use Transparency 7 to show students how to do calculations with time. Then have students apply the INT÷ key in the student activity *It's About Time!*

Five-Minute Fillers

1. Can you predict the series of repeating digits in the answer when you divide the following numbers by 9?

   2, 5, 8, 7, 12, 16, 32, 71, 43
2. Can you predict the series of repeating digits in the answer when you divide the following numbers by 99?

   6, 8, 23, 49, 88, 44, 125, 142, 164, 189

It's About Time!

Directions

Use the INT÷ key to convert the following units of time.

a. 300 seconds = ______ minutes ______ seconds

b. 1234 seconds= ______ minutes ______ seconds

c. 65 seconds = ______ minutes ______ seconds

d. 1845 seconds= ______ minutes ______ seconds

e. 3600 seconds= ______ minutes ______ seconds

f. 125 seconds = ______ minutes ______ seconds

g. 300 minutes = ______ hours ______ minutes

h. 75 minutes = ______ hours ______ minutes

i. 320 minutes = ______ hours ______ minutes

j. 3600 minutes = ______ hours ______ minutes

k. 1605 minutes = ______ hours ______ minutes

l. 131 minutes = ______ hours ______ minutes

m. 77 hours = ______ days ______ hours

n. 168 hours = ______ days ______ hours

o. 332 hours = ______ weeks ______ days ______ hours

p. 89 days = ______ weeks ______ days

q. 1024 days = ______ years ______ days

r. 2231 days = ______ years ______ weeks ______ days

s. 186 months = ______ years ______ months

Whole-Number Division

Directions

Use the fractions below:

1. Use the [INT÷] key to find the quotient and the remainder for the problem.
2. Estimate the decimal answer to the nearest hundredth for each answer.
3. To check your answer, use the [÷] key.

Problem	Quotient, Remainder	Decimal Estimate	Calculator Solution
a. 400 ÷ 4	________	________	________
b. 405 ÷ 4	________	________	________
c. 4050 ÷ 4	________	________	________
d. 875 ÷ 14	________	________	________
e. 8755 ÷ 14	________	________	________
f. 37,146 ÷ 14	________	________	________
g. 179 ÷ 33	________	________	________
h. 9814 ÷ 65	________	________	________
i. 58,173 ÷ 65	________	________	________
j. 275 ÷ 132	________	________	________

True to Form

Directions

1. To get a decimal, divide using the [÷] key.
2. If there is a decimal portion of the answer, you can get a whole-number remainder by following these steps:
   a. Enter the dividend into the display of the Math Explorer.
   b. Press the [INT÷] key.
   c. Enter the divisor.
   d. Press the [=] key.
3. Express the answer to each problem in decimal and in remainder form.

Problem	Answer in decimal form	Answer in remainder form
a. 494 ÷ 98	5.0408163	5 R 4
b. 7035 ÷ 552		
c. 4805 ÷ 866		
d. 1722 ÷ 631		
e. 316,493 ÷ 1542		
f. 8442 ÷ 8199		
g. 850,091 ÷ 2691		
h. 87,808 ÷ 760		
i. 11,428 ÷ 143		

Measuring Up and Down

Directions

Use the [INT÷] key or the [×] key to convert the following units.

a. 100 inches = ________ feet ________ inches

b. 79 feet = ________ yards ________ feet

c. 17 feet = ________ inches

d. 185 ounces = ________ pounds ________ ounces

e. 13 pounds 5 ounces = ________ ounces

f. 172 eggs = ________ dozen + ________ eggs

g. 12,000 yards = ________ miles ________ yards

h. 7.7 miles = ________ feet

i. 85 quarts = ________ gallon ________ quarts

j. 56 quarts = ________ cups

Can you work these more challenging problems?

k. 156 square inches = ________ square feet ________ square inches

l. 108 square feet = ________ square yards ________ square feet

m. 3.5 square feet = ________ square inches

n. 3 cubic feet = ________ cubic inches

o. 700 cubic feet = ________ cubic yards ________ cubic feet

p. If 231 cubic inches = 1 gallon, 5 cubic feet = ________ gallons, ________ cubic inches.

Rules Rule!

Directions

1. Discover the rule for each set of problems.
2. Express each amount in the usually preferred manner.
3. Label your answer appropriately.

6 days — 1 wk 5 da 2 qt — 1 gal 53 min — 1 hr ____ min 5 days — ____ wk ____ da 10 in. — ____ ft ____ in. Rule: ____________	40 min — 2 hr 2 ft — 2 yd 3 c — ____ qt ____ c $0.43 — $____ 3 wk — ____ mo ____ wk Rule: ____________
1 ft 8 in. — 5 in. 1 hr 20 min — 20 min $2.40 — ____________ 2 wk 6 da — ____________ 3 doz eggs — ____________ 1 yr — ____________ Rule: ____________	1 pt — 1 gal 8 oz — 4 lb 18 hr — ____________ 1 c — ____________ 10 in. — ____________ 59 min — ____________ Rule: ____________

Lesson 7: Number Properties

Overview

Students will use the calculator to help them learn and apply the number properties of multiplication and addition to simplify problems.

Transparencies

Use Transparency 26, *Parentheses*, Chapter 2

Keys Introduced

None

Student Worksheets

What is Your Group?
Family Friends?
Workout!
Subtraction and Division Order
Mind over Calculator

Teaching Steps

1. Use Transparency 26, *Parentheses*, to help students understand that parentheses mean "Me first!"
2. Have some of the students try to solve the worksheet problems mentally while others are working the problems on the Math Explorer.
3. The definitions of the number properties students need to understand are:

   Associative property:

   $a + (b + c) = (a + b) + c$

   $a \times (b \times c) = (a \times b) \times c$

   Commutative property:

   $a + b = b + a$

   $a \times b = b \times a$

Distributive property:

$a \times (b + c) = a \times b + a \times c$

$(b + c) \times a = b \times a + c \times a$

Five-Minute Filler

Write down any eight-digit number. Enter that number into the display of the Math Explorer using only the [1] [0], [+] and [=] keys.

What Is Your Group?

Directions

1. Work the following problems using the parenthesis keys on the Math Explorer when indicated.
2. Write a rule you think applies to this type of problem.

   ____

3. Can you use this rule to solve the problems mentally as fast as you did with the Math Explorer?

a. $(155 + 45) + 692$ ________

$155 + 45 + 692$ ________

$155 + (45 + 692)$ ________

b. $(567 + 133) + 904$ ________

$567 + 133 + 904$ ________

$567 + (133 + 904)$ ________

c. $(318 + 682) + 756$ ________

$318 + 682 + 756$ ________

$318 + (682 + 756)$ ________

d. $(432 + 568) + 969$ ________

$432 + 568 + 969$ ________

$432 + (568 + 969)$ ________

e. $(5 \times 20) \times 127$ ________

$5 \times 20 \times 127$ ________

$5 \times (20 \times 127)$ ________

f. $(40 \times 5) \times 78$ ________

$40 \times 5 \times 78$ ________

$40 \times (5 \times 78)$ ________

g. $(15 \times 10) \times 30$ ________

$15 \times 10 \times 30$ ________

$15 \times (10 \times 30)$ ________

h. $(40 \times 25) \times 72$ ________

$40 \times 25 \times 72$ ________

$40 \times (25 \times 72)$ ________

Family Friends?

Directions

1. Work the following problems using the Math Explorer.

   a. $45 + 67 + 55$ ________

   $67 + 55 + 45$ ________

   $45 + 55 + 67$ ________

   b. $12.5 \times 8 \times 37.5$ ________

   $37.5 \times 8 \times 12.5$ ________

   $8 \times 37.5 \times 12.5$ ________

   c.

29	29	16	46	58	37	58
37	16	29	37	37	46	16
46	58	58	58	46	58	29
58	46	37	29	29	16	37
+16	37	46	16	16	29	46

   d. $5 \times 6 \times 7 \times 8 \times 9$ ________

   $6 \times 7 \times 8 \times 9 \times 5$ ________

   $7 \times 8 \times 9 \times 5 \times 6$ ________

   $8 \times 9 \times 5 \times 6 \times 7$ ________

   $9 \times 5 \times 6 \times 7 \times 8$ ________

2. Can you write a rule for working problems in which the commutative property holds?

____

____

Subtraction and Division Order

Directions

1. Use the Math Explorer to find the answers to the following problems.
2. Describe the difference in the two answers for each set of problems.
3. Is subtraction commutative? _____

   Is division commutative? _____

a. 12 - 5 ______

   5 - 12 ______

b. 36 ÷ 6 ______

   6 ÷ 36 ______

c. 24 - 8 - 3 ______

   8 - 3 - 24 ______

d. 8 ÷ 2 ______

   2 ÷ 8 ______

e. 14 - 6 ______

   6 - 14 ______

f. 21 ÷ 3 ______

   3 ÷ 21 ______

g. 27 - 19 ______

   19 - 27 ______

h. 12 ÷ 3 ______

   3 ÷ 12 ______

Workout!

When you exercise, you need to work hard enough so that your heart gets a workout but not so hard that you overdo it. Exercise too hard, and you get exhausted. Don't exercise hard enough, and you won't get the benefit you want. Your heart rate per minute indicates how hard you are exercising.

The lower limit that your heart rate (beats per minute) should be during exercise is called your minimum aerobic heart rate. The upper limit that your heart rate should be during exercise is called your maximum aerobic heart rate. When you exercise, you want your heart rate to be somewhere between your minimum and maximum.

Here's how to calculate your minimum and maximum aerobic heart rates (these formulas are estimates for any age and for either sex):

* Minimum aerobic heart rate: (220 - age) × 0.7

* Maximum aerobic heart rate: (220 - age) × 0.85

Example: Susan is 15 years old.

(220 - 15) × 0.7 = 143.5 = Susan's minimal aerobic heart rate

(220 - 15) × 0.85 = 174.25 = Susan's maximal aerobic heart rate

So, when Susan exercises, she should try to keep her heart rate between 143 and 174 beats per minute.

Use the parenthesis keys on your Math Explorer to calculate your minimum and maximum aerobic heart rates. Then calculate the minimum and maximum heart rates for some of your classmates, brothers or sisters, parents, and so on.

Name	Age	Min. Heart Rate (220 - age) × .7	Max.l Heart Rate (220 - age) × .85
______	______	____________	____________
______	______	____________	____________
______	______	____________	____________

Mind over Calculator

Directions

1. Work the following problems using the Math Explorer.

   a. $83 \times 57 + 83 \times 43$ _____ $83 \times (57 + 43)$ _____

   b. $456 \times 18 + 544 \times 18$ _____ $(456 + 544) \times 18$ _____

   c. $96 \times 47 + 4 \times 47$ _____ $(96 + 4) \times 47$ _____

   d. $861 \times 366 + 861 \times 634$ _____ $861 \times (366 + 634)$ _____

   e. $204 \times 13 + 17 \times 204$ _____ $204 \times (? + ?)$ _____

2. Solve these problems using the Math Explorer; try to predict the answer.

   a. 30×30 _______ $(40 + 0) \times (40 - 0)$ _______

   b. 31×29 _______ $(40 + 1) \times (40 - 1)$ _______

   c. 32×28 _______ $(40 + 2) \times (40 - 2)$ _______

   d. 33×27 _______ $(40 + 3) \times (40 - 3)$ _______

   e. 34×26 _______ $(40 + 4) \times (40 - 4)$ _______

   f. 35×25 _______ $(40 + 5) \times (40 - 5)$ _______

3. In your own words, describe the distributive property.

   ____

Lesson 8: Selecting the Number of Decimal Places

Overview

Students learn how to use the Fix key to select the number of places after the decimal point.

Transparencies

Transparency 16, *Fixing the Decimal Point*, Chapter 2

Keys Introduced

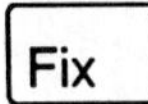

Student Worksheets

Five Alone
Endless Fractions?
Formula Four
Taxing Problems

Teaching Steps

1. Use Transparency 16, *Fixing the Decimal Point*, to demonstrate how to set the number of places after the decimal point.
2. Have students practice setting various decimal places from 0 to 7 places.
3. Explain:
   - The decimal place can be fixed any time during a calculation.
   - The Math Explorer automatically rounds up or down to the number of decimal places selected:

Press	Display
. 147	0.147
Fix 2	0.15

Note that when the decimal is set to display two places, the calculator rounds 0.147 up to 0.15. Try Fix 1. What happens?

4. As a group, have students practice rounding decimals on the calculator:
   - Have the students fix the decimal to 1.
   - Then have them enter a decimal on the calculator. Tell them not to press the [=] key yet.
   - Next, have them estimate what the decimal would be if it were rounded to tenths.
   - Once they have made their estimates, tell them to press [=].
5. Show students they can clear any [Fix] key settings by pressing [Fix], [.], or [ON/AC]. Students can then enter numbers in a floating-decimal format.

Five-Minute Filler

Historical note: Simon Stevin, a Dutchman, wrote a book in 1585 called *La Disme*. In this book he described the use of decimals and gave the rules for computing with them. During the sixteenth and seventeenth centuries, several symbols for decimals were used. Here are some examples of how 3.14 was written during this time.

3/14 $3/\underline{14}$ 314...(2)   3 (0) 1 (1) 4 (2)

3 (14) $3_{0}14$ $3,1^{1}4^{12}$

Write a number involving a decimal for each measurement:

a. Money ________

b. Temperature ________

c. Circumference of a Circle ________

Five Alone

Directions

1. Use the Math Explorer to find a number halfway between 0.4865 and 0.512 by adding these two decimals and dividing by 2. (Use the [Fix] key to round your answer to three decimal places.)
   a. Add that answer to 0.512 and divide by 2 again.
   b. Repeat that process five times.
   c. As you repeat this process you get closer and closer to what number?

   ____

2. Use the Math Explorer to find a number halfway between 0.0715 and 0.1982 by adding these two decimals and dividing by 2. (Use the [Fix] key to round your answer to four decimal places.)
   a. Add that answer to 0.1982 and divide by 2 again.
   b. How many times do you have to add 0.1982 and divide by 2 before you get one of the original numbers?

   ____

Endless Fractions?

Directions

1. Find the decimal representation for each fraction by entering the fraction into the display of the Math Explorer and pressing the [F↔D] key.
2. Circle the fractions that appear to yield repeating decimals rather than terminating decimals.
3. How did the fractions you circled differ from the ones you didn't circle?

a. $\frac{7}{12}$ ____________

b. $\frac{1}{8}$ ____________

c. $\frac{5}{6}$ ____________

d. $\frac{3}{7}$ ____________

e. $\frac{11}{20}$ ____________

f. $\frac{6}{25}$ ____________

Formula Four

Directions

Solve the following problems using justifiable numbers of decimal places in your answers. Use the [Fix] key to set the number of places.

1. The deep-space probe Pioneer 10 took 21 months to get from Mars to Jupiter, a distance of 998 million kilometers. Assume 30.5 days per month. Find the probe's average speed in kilometers per hour during that period.

   ____

$$\text{Average speed} = \frac{\text{distance}}{\text{time}}$$

2. Kurt wanted to pour a concrete driveway from his garage to the street. The driveway will be 23.5 feet long, 12.6 feet wide and .5 feet thick. How many cubic yards of concrete should he order? (Hint: There are 27 cubic feet in a cubic yard.)

   ____

3. In the first 6 months of 1986, Canada drilled 1457 new oil wells with a total depth of 5.4 million feet. What is the average depth of each of these wells?

   ____

4. Elena travels the length of the Pennsylvania Turnpike (360 miles) in 7 hours. What is her average speed in miles per hour?

   ____

Taxing Problems

Directions

Solve the following problems, leaving your answers with a justifiable number of decimal places. Use the [Fix] key to set the number of decimal places.

1. A job that pays $7.78 an hour yields a gross income of $311.20 in a 40-hour week. Compute the deductions and determine the net pay.

   a. Federal income tax on $311.20 for a single person is 0.245 times that amount. Federal income tax ______

   b. State income tax is .037 × $311.20. State income tax ______

   c. FICA, or social security tax, is 0.0745 times $311.20. FICA tax ______

   d. Net pay is gross pay minus deductions. Net pay ______

2. The basic unit for measuring electricty is the kilowatt hour (kWh). This is the amount of electrical energy required to operate a 1000-watt appliance for 1 hour. The table shown contains the average number of kilowatt-hours for operating the appliances for 1 month.

   a. What is the sum of the kilowatt-hours for operating these seven appliances for one month? Record this at the bottom of the chart.

   b. At a cost of $0.055 for each kilowatt-hour, what is the monthly cost for operating each of these appliances?

Appliance	kWh/ month	Cost/ month
Micro-wave	15.8	_____
Range	97.6	_____
Refrig-erator	152.4	_____
Water heater	400.0	_____
Radio	7.5	_____
T.V.	55.9	_____
Air Con-ditioner	1210	_____
Totals	_____	_____

Lesson 9: Using the Memory Keys

Overview

Students learn how to use the memory keys to store values in the memory, add numbers to the memory, subtract numbers from the memory, and recall the values from the memory.

Transparencies

Transparency 11, *Memory Keys*, Chapter 2

Keys Introduced

Student Worksheets

Step by Step
Having a Party
Separated
In the Shade

Teaching Steps

1. Use Transparency 11, *Memory Keys*, to demonstrate how to use the memory keys.
2. Demonstrate how the [x⇄M] key exchanges the value in the display with the value in the memory.
3. Demonstrate how using the [M+] or [M–] key adds or subtracts the value in the display from the value in the memory. Ask students to explain the difference in function between these two keys and the [x⇄M] key.
4. If appropriate, tell students that fractions, decimals, and negative numbers can be stored in memory. Operations also can be done on those numbers while they are in the memory.
5. Have students complete the worksheet, *Having a Party*.

Five-Minute Filler

I bought 25 pencils at 13¢ each, and 7 pads of paper at $1.29 each, but returned 5 erasers I had purchased earlier for 35¢ each. What was my expense?

Step by Step

Directions

1. Identify the first part of the problem, solve it, and enter the answer in the memory by pressing the [x⊂M] key.

2. Solve the second part of the problem. Press the necessary operation key, then the [MR] key and then the [=] key.

   a. Andrea bought 5 pounds of hamburger at $1.29 per pound. She also bought 4 packages of buns for $0.69 per package. What was the cost of her purchases?

   ____

   b. Tom needed some supplies for a tune-up he planned on his car. He bought 5 quarts of oil at $1.39 per quart and 6 spark plugs at $1.89 per plug. How much did he spend?

   ____

   c. Mr. Munro had 23.5% of his gross annual salary of $42,568 withheld for federal income tax. He had 4.1% of his gross salary withheld for state income tax. How much total tax was withheld?

   ____

   d. Ellen had a $2500 time savings certificate that paid 7.25% interest. Her savings account paid 5.75% interest on the $1500 she had deposited in it. How much interest would she earn in 1 year from these two accounts?

   ____

Having a Party

Directions

You and your friends are going to have a picnic. You are in charge of food. Use the memory keys to calculate what the total bill will be.

Item	Quantity	Cost per item	Total
Luncheon meats	3 packages	$1.19	______
Apple cider	2 gallons	$1.59	______
Bread	2 loaves	$0.89	______
Dried fruit	4 bags	$1.39	______

The total cost is: $ __________.

Separated

Directions

1. Divide the following problems into two separate parts.
2. Enter the answer for the part requiring the first operation into the Math Explorer by pressing the [x⊂M] key.
3. After the second answer is obtained, perform the indicated operation on that number with the number in the memory of the Math Explorer.

a. $5.791 - \frac{3.462}{32.576}$ ____________

b. $32.576 - \frac{5.791}{3.462}$ ____________

c. $\frac{56.74}{4.95 - 2.81}$ ____________

d. $56.74 - \frac{4.95}{2.81}$ ____________

e. $36 + 54(1.04)^4$ ____________

f. $7^4 - 9^3$ ____________

In the Shade

Directions

1. Find the area of the shaded part of each figure.
2. Use the formulas $C=\pi D$ and $A=\pi r^2$.

$r = 5.4$ cm

$A =$ ________

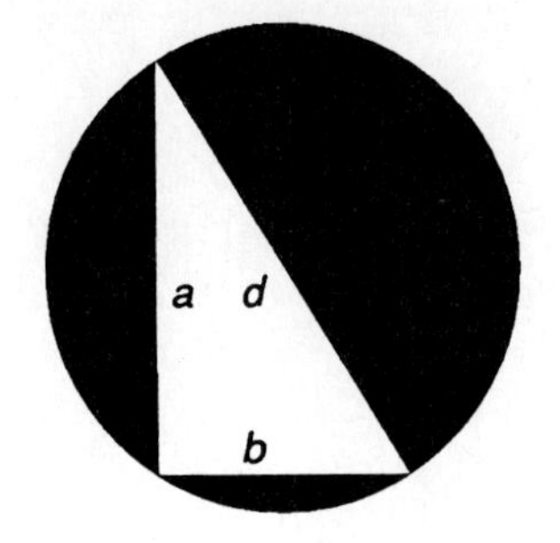

$a = 12$ cm, $b = 5$ cm, $d = 13$ cm,

$A =$ ________

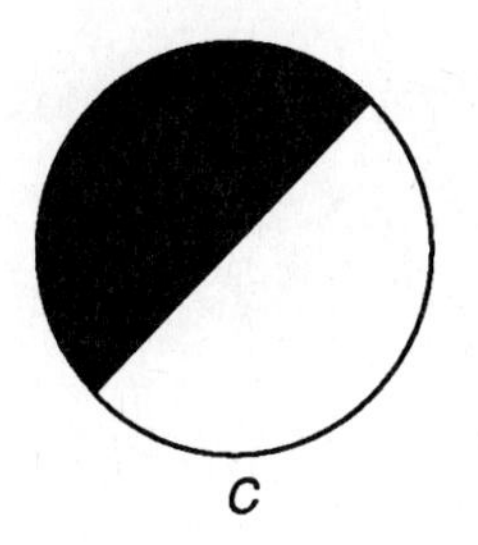

$C = 47.1$ cm

$A =$ ________

Lesson 10: Simplifying Fractions

Overview

Students learn how to use the calculator to simplify fractions in two ways: 1) the calculator chooses the common factor and 2) students choose the common factor.

Transparencies

Transparencies 17 and 18, *Simplifying Fractions: You Choose a Common Factor* and *Simplifying Fractions: Calculator Chooses a Factor*, Chapter 2

Keys Introduced

Student Worksheets

How Low Can You Go?
Is it Simple?
Going Up?
Is it the GCF, the GED, or the SOP?

Teaching Steps

Calculator Chooses a Common Factor

1. Use Transparency 18, *Simplifying Fractions: Calculator Chooses a Factor*, to show students how to enter fractions on the calculator and how to simplify them.
2. Be sure to point out:
   - Pressing the [Unit] key enters the whole number portion of a mixed number.
   - Pressing the [/] key separates the numerator of the fractions from the denominator.
   - The indicator **u** in the display separates whole numbers from fractions.
   - If students use the wrong key sequence to enter a fraction, **Error F** appears in the display.

Press	**Display**
2 [/] [=]	Error F

- The maximum number of digits that can be entered for the numerator and denominator is eight; for the denominator only, it is three; and for a mixed number, it is seven.

3. Have students enter 2/4 on the calculator and press [Simp].
   - Point out that **SIMP** always appears in the display when they press the [Simp] key.
   - **N/D → n/d** also appears in the display if the fractions can be simplified.
4. Have students press [=] to complete the simplification of 2/4. Point out that the calculator reduces the fraction by using the smallest common factor, so simplification may take several steps.
5. Tell students to press [x⇄y] anytime they want to see the factor that the calculator used. Have them press [x⇄y] again to return to the fraction.
6. Have students practice simplifying these fractions:

   a. $\frac{6}{20}$ b. $\frac{4}{16}$ c. $\frac{2}{3}$ d. $\frac{15}{30}$

Student Chooses a Common Factor

1. Use Transparency 17, *Simplifying Fractions: You Choose a Common Factor*, to show students how they can choose the factor.
2. Have students practice simplifying these fractions:

   a. $\frac{16}{32}$ b. $\frac{6}{20}$ c. $\frac{1}{5}$ d. $\frac{23}{69}$
3. Summarize the steps for the two methods of simplification:
   - Calculator chooses a factor:

     Enter fraction, press [Simp] [=].
   - Students choose a factor:

   Enter fraction, press [Simp], enter factor, press [=].
4. Have students complete the worksheet, *How Low Can You Go?*

Five-Minute Filler

Choose any three-digit number in which the first digit is larger than the last. Reverse the digits and subtract. Repeat with another three-digit number. Express your two differences as a fraction. Can it be reduced? Can it always be reduced?

How Low Can You Go?

Directions

Each of the fractions in the chart below can be simplified. Your challenge is to find the greatest common factor that will reduce the fraction to simplest terms in your first try.

1. Enter the fraction from the chart into the calculator.
2. Estimate the greatest common factor, and write your estimate on the chart below.
3. When you have written your estimate on the chart:
   - Press [Simp]. Enter your estimate of the greatest common factor. Press [=].
   - If **N/D → n/d** does not appear in the display, the factor you chose was the greatest common factor. Write 3 points on the chart below. Go to the next fraction.
   - If **N/D → n/d** does appear in the display, the factor you chose was not the greatest common factor. Try again. If you make it on your second try, write 2 points on the chart below. If you make the correct guess on your third try, give yourself 1 point. After three tries, you do not receive any points.

Fraction	First Guess	Second Guess	Third Guess	Points
a. $\frac{12}{20}$	______	______	______	______
b. $\frac{15}{25}$	______	______	______	______
c. $\frac{4}{12}$	______	______	______	______
d. $\frac{30}{45}$	______	______	______	______
e. $\frac{21}{24}$	______	______	______	______

f. $\frac{5}{40}$ ________ ________ ________ ________

g. $\frac{48}{144}$ ________ ________ ________ ________

h. $\frac{16}{9}$ ________ ________ ________ ________

i. $\frac{27}{81}$ ________ ________ ________ ________

j. $\frac{16}{48}$ ________ ________ ________ ________

k. $\frac{8}{24}$ ________ ________ ________ ________

l. $\frac{21}{56}$ ________ ________ ________ ________

Total score ____________

Is It Simple?

Directions

1. Circle the fractions you guess are in simplest form.
2. Enter each fraction into the display of the Math Explorer.
3. Press the [Simp] key and then the [=] key. If the fraction doesn't change, it is in simplest form. Record your answer.

a. $\frac{17}{51}$ ________

b. $\frac{32}{43}$ ________

c. $\frac{33}{54}$ ________

d. $\frac{36}{93}$ ________

e. $\frac{78}{91}$ ________

f. $\frac{14}{39}$ ________

g. $\frac{13}{52}$ ________

h. $\frac{27}{72}$ ________

i. $\frac{15}{96}$ ________

j. $\frac{15}{51}$ ________

k. $\frac{23}{92}$ ________

l. $\frac{15}{59}$ ________

m. $\frac{81}{108}$ ________

n. $\frac{63}{89}$ ________

o. $\frac{54}{81}$ ________

p. $\frac{26}{39}$ ________

q. $\frac{56}{72}$ ________

r. $\frac{9}{16}$ ________

s. $\frac{7}{45}$ ________

t. $\frac{19}{57}$ ________

Going Up?

Directions

1. Changing fractions to higher terms simply means that both the numerator and denominator are multiplied by the same number.

Example: Express $\frac{2}{5}$ in 20ths.

Step 1. Divide the original denominator into the new one.

$20 \div 5 = 4$

Step 2. Multiply the numerator by the answer.

$2 \times 4 = 8$

Therefore, $\frac{2}{5} = \frac{8}{20}$

Step 3. Check your answer by reducing the new fraction on the Math Explorer to see if you arrive at the original fraction at some step in the reduction process.

2. Find the equivalent fraction with the indicated denominator:

a. $\frac{4}{5}$ to thirtieths ___

b. $\frac{15}{16}$ to sixty-fourths ___

c. $\frac{4}{5}$ to fortieths ___

d. $\frac{3}{11}$ to seventy-sevenths ___

e. $\frac{7}{12}$ to forty-eighths ___

f. $\frac{10}{12}$ to seventy-seconds ___

g. $\frac{1}{3}$ to forty-fifths ___

h. $\frac{15}{24}$ to ninety-sixths ___

i. $\frac{11}{13} = \frac{\quad}{169}$

j. $\frac{12}{14} = \frac{\quad}{196}$

k. $\frac{13}{17} = \frac{\quad}{187}$

l. $\frac{5}{11} = \frac{\quad}{121}$

m. $\frac{16}{32} = \frac{\quad}{1280}$

n. $\frac{17}{23} = \frac{\quad}{161}$

o. $\frac{11}{24} = \frac{\quad}{528}$

p. $\frac{5}{23} = \frac{\quad}{115}$

Is It the GCF, the GED, or the SOP?

Directions

1. Enter the given fraction into the Math Explorer.
2. Press the [Simp] key and then the [=] key.
3. Look at the display, if **N/D → n/d** appears in the display, the fraction can be simplified further.
4. Press the [x⇄y] key. This tells you the factor the Math Explorer used in reducing the fraction.
5. Press the [x⇄y] key again. This will return you to the fraction.
6. Continue the process until the **N/D → n/d** symbol no longer appears in the display.
7. The product of the factors used by the Math Explorer is the greatest common factor of the numerator and denominator of the original fraction.
8. Simplify the fractions and identify the GCF.

a. $\frac{14}{40}$ ______ i. $\frac{88}{165}$ ______

b. $\frac{20}{64}$ ______ j. $\frac{25}{36}$ ______

c. $\frac{27}{48}$ ______ k. $\frac{24}{64}$ ______

d. $\frac{21}{24}$ ______ l. $\frac{27}{72}$ ______

e. $\frac{81}{144}$ ______ m. $\frac{50}{100}$ ______

f. $\frac{24}{32}$ ______ n. $\frac{156}{144}$ ______

g. $\frac{25}{125}$ ______ o. $\frac{91}{343}$ ______

h. $\frac{18}{32}$ ______ p. $\frac{117}{182}$ ______

Lesson 11: Greatest Common Factor and Least Common Multiple

Overview

Students learn how to use the calculator to find the GCF and LCM.

Transparencies

Use Transparencies 17 and 18, *Simplifying Fractions: You Choose a Common Factor* and *Simplifying Fractions: Calculator Chooses a Factor*, Chapter 2

Keys Introduced

None

Student Worksheets

Common Factors
LCM
Euclid's Secret
Eighth-grade Discovery
GCF

Teaching Steps GCF & LCM:

1. Use Transparencies 17 and 18, *Simplifying Fractions*, to review finding factors in fractions.
2. Review the Lesson 10 Worksheet, *Is It the GFC, the GED, or the SOP?*
3. Work through the directions for each of the activities in this lesson with the students.
4. Check the first few answers the students get for each of the activities.

Five-Minute Filler

Suppose we want to tile a floor that is 15 feet by 18 feet. Only square tiles are available, in sizes 4, 5, 8, or 9 inches on a side. What size should we buy if we do not want to cut any tiles?

Common Factors

Directions

The greatest common factor (GCF) is the product of the common prime factors of each of the numbers being considered. As the Math Explorer reduces fractions, it divides both the numerator and the denominator by a common prime number.

1. Enter each fraction into the display of the Math Explorer.
2. Press the [Simp] key and then the [=] key.
3. Press the [x⇄y] key. This will give you the common prime number the calculator used.
4. Record the factor.
5. Repeat steps 2-4 until the fraction cannot be reduced further. Find the product of all the factors.
6. Find the GCF and the reduced fraction for each given fraction.

	GCF	Reduced Fraction
a. $\frac{121}{165}$	____________	____________
b. $\frac{144}{156}$	____________	____________
c. $\frac{480}{512}$	____________	____________
d. $\frac{405}{729}$	____________	____________
e. $\frac{128}{640}$	____________	____________
f. $\frac{343}{735}$	____________	____________
g. $\frac{236}{944}$	____________	____________
h. $\frac{242}{528}$	____________	____________

LCM

Directions

1. When the Math Explorer adds or subtracts fractions with different denominators, it selects the least common multiple of the two denominators to use as a common denominator.

   a. Continue the multiples of:

   5 10 15 20 _____ _____ _____ _____ _____

   7 14 21 28 _____ _____ _____ _____ _____

   b. Find at least two common multiples. Underscore the least common multiple. This LCM becomes the denominator for the answer when two fractions with denominators 5 and 7 are added or subtracted.

2. Another way to find the LCM is to follow these steps.

   a. Find the prime factors of each number. For example, to find the LCM of 147 and 163, write

   $147 = 3 \times 7 \times 7$ $63 = 3 \times 3 \times 7$

   b. Express the factors as exponents.

   $147 = 3 \times 7^2$ $63 = 3^2 \times 7$

   c. Pick the largest power of every factor in the numbers.

   3^2 and 7^2

   d. The product of the largest powers of these factors is the least common multiple.

   $3^2 \times 7^2 = 9 \times 49 = 441$

3. Use your calculator and the second procedure to find the LCM of the following pairs of numbers. Match each letter with an answer.

a. 26, 39 _______ (1). 72

b. 16, 20 _______ (2). 245

c. 35, 49 _______ (3). 80

d. 18, 21 _______ (4). 270

e. 45, 54 _______ (5). 78

(6). 126

Euclid's Secret

Directions

1. Using the Math Explorer, divide the larger number of each pair by the smaller.
2. If the remainder is zero, the GCF is the smaller number.
3. If the remainder is not zero, divide the previous divisor by the remainder.
4. If the remainder this time is zero, the GCF is the first remainder.
5. If the remainder is not zero, divide the first remainder by the second.
6. Continue in this manner until a remainder of zero is obtained.
7. The last divisor is the GCF.
8. Find the GCF of the following pairs of numbers.

Example:

```
      2
 81)216
    162   1
     54)81
        54   2
        27)54
           54
```

The GCF of 81 and 216 is 27.

a. 130, 265 ________

b. 336, 392 ________

c. 270, 315 ________

d. 124, 158 ________

e. 99, 231 ________

f. 3551, 4489 ________

Eighth-Grade Discovery

Directions

1. Andrew Nathan, an eighth-grade mathematics student, made the following observation:

   a. Take a fraction that is not in its simplest form, and subtract the numerator from the denominator.

   b. Simplify the fraction, and then subtract the numerator from the denominator.

   c. Divide the result obtained in the first step by the result obtained in the second step.

   d. The quotient obtained in the third step seems to be the greatest common factor (GCF) of the numerator and denominator of the given fraction.

   For example, given the fraction $\frac{35}{45}$, steps **a** through **c** would yield the following results:

   Step a. 45 - 35 = 10

   Step b. 35 ÷ 45 = 7 ÷ 9 and 9 - 7 = 2

   Step c. 10 ÷ 2 = 5

   The GCF of 35 and 45 is 5.

2. Andrew surmised that this would be true for any fraction.

   Enter each fraction into the Math Explorer. Press the [Simp] and [=] keys. To check the factor the Math Explorer used, press the [x⇆y] key. If you must simplify a second or third time, multiply the factors to find the greatest common factor. Compare your answers to those you found using Andrew's conjecture.

   To test if your answer is the greatest common factor, divide both the numerator and denominator by the GCF you obtained. Does this produce a completely simplifed fraction?

Test Andrew's conjecture on the following fractions.

a. $\frac{55}{90}$ ______

d. $\frac{165}{198}$ ______

g. $\frac{84}{144}$ ______

b. $\frac{30}{42}$ ______

e. $\frac{56}{96}$ ______

h. $\frac{64}{92}$ ______

c. $\frac{168}{280}$ ______

f. $\frac{91}{169}$ ______

i. $\frac{156}{220}$ ______

GCF

Directions

Each of the fractions in the chart below can be simplified. Your challenge is to find the greatest common factor that will reduce the fraction to simplest terms in your first try.

1. Enter the fraction from the chart into the Math Explorer.
2. Estimate the greatest common factor, and write your estimate on the chart below.
3. When you have written your estimate on the chart:
   - Press [Simp]. Enter your estimate of the GCF. Press the [=] key.
   - If **N/D** → **n/d** does not appear on the display, the factor you chose was the greatest common factor. Write 3 points on the chart below. Go to the next fraction.
   - If **N/D** → **n/d** does appear in the display, the number you chose was not the greatest common factor. Try again. If you make it on your second try, write 2 points on the chart below. If you make a correct guess on your third try, give yourself 1 point. After three tries, you do not receive any points.

Fraction	First Guess	Second Guess	Third Guess	Points
a. $\frac{8}{4}$	______	______	______	______
b. $\frac{10}{5}$	______	______	______	______
c. $\frac{2}{8}$	______	______	______	______
d. $\frac{6}{90}$	______	______	______	______
e. $\frac{21}{45}$	______	______	______	______

f. $\frac{22}{40}$ ________ ________ ________ ________

g. $\frac{48}{96}$ ________ ________ ________ ________

h. $\frac{12}{144}$ ________ ________ ________ ________

i. $\frac{21}{27}$ ________ ________ ________ ________

j. $\frac{84}{144}$ ________ ________ ________ ________

k. $\frac{18}{93}$ ________ ________ ________ ________

l. $\frac{136}{145}$ ________ ________ ________ ________

Total score ____________

Lesson 12: Operations on Fractions

Overview

This lesson teaches students how to perform operations on fractions.

Transparency

Transparency 20, *Math Operations: Fractions*, Chapter 2

Keys Used

Student Worksheets

Fraction Estimation
Add and Subtract Fractions
Multiply and Divide Fractions
Rule of Thumb
Rule of Thumb, Jr.
Best Guess
Good, Better, Best
Tables of Stone?

Teaching Steps

1. Use Transparency 20, *Math Operations: Fractions*, to review adding and subtracting fractions.
2. Explain that when two fractions are added or subtracted, the calculator will express and display the answer with the lowest common denominator.

   Addition Example:

Press	Display
1 [/] 2 [+]	1/2
3 [/] 8 [=]	7/8

   Note: 8 is the lowest common denominator of 2 and 8.

Subtraction Example:

Press	Display
7 [/] 12 [-]	7/12
5 [/] 9 [=]	1/36

Note: 36 is the lowest common denominator of 12 and 9.

3. If appropriate, show students that if fractions and decimals are added or subtracted, the answer will be displayed as a decimal. Use [FOD] to convert the decimal result to a fraction.

Example:

Press	Display
1 [/] 2 [+]	1/2
1 [.] 6 [=]	2.1
[FOD]	2u 1/10

Note: If a decimal result has four or more digits to the right of the decimal point, you cannot convert the result to a fraction due to display limitations.

4. Have students complete the worksheets, *Add and Subtract Fractions* and *Multiply and Divide Fractions.*

Five-Minute Filler

John has $\frac{7}{8}$ inches of platinum wire that he must use to repair transistors. If he uses $\frac{3}{32}$ inches of wire on each transistor, how many can he repair with the wire he has?

Fraction Estimation

Directions

Estimate and match the answer you think is best for each. Use Math Explorer to verify your estimation.

Estimation 1 Addition of Fractions

1. $\frac{1}{2} + \frac{1}{2} =$ 2. $\frac{1}{3} + \frac{1}{2} =$ 3. $\frac{1}{5} + \frac{1}{6} =$ 4. $\frac{1}{8} + \frac{1}{4} =$

______ ______ ______ ______

Answers: a. $\frac{1}{3}$ b. $\frac{1}{4}$ c. 1 d. $\frac{1}{2}$ e. 2

Estimation 2 Addition of Fractions

1. $\frac{1}{2} + \frac{2}{3} =$ 2. $\frac{3}{4} + \frac{2}{3} =$ 3. $\frac{4}{5} + \frac{1}{3} =$ 4. $\frac{3}{8} + \frac{4}{8} =$

______ ______ ______ ______

Answers: a. 1 b. $\frac{3}{2}$ c. 2 d. $\frac{3}{4}$ e. $\frac{1}{2}$

Estimation 3 Subtraction of Fractions

1. $\frac{1}{2} - \frac{1}{3} =$ 2. $\frac{1}{5} - \frac{1}{6} =$ 3. $\frac{1}{4} - \frac{1}{8} =$ 4. $\frac{1}{2} - \frac{1}{4} =$

______ ______ ______ ______

Answers: a. $\frac{1}{2}$ b. $\frac{1}{4}$ c. $\frac{1}{3}$ d. $\frac{1}{8}$ e. 1

Estimation 4 Subtraction of Fractions

1. $\frac{2}{3} - \frac{1}{3} =$ 2. $\frac{3}{4} - \frac{1}{2} =$ 3. $\frac{7}{8} - \frac{2}{3} =$ 4. $\frac{3}{5} - \frac{1}{4} =$

______ ______ ______ ______

Answers: a. $\frac{1}{4}$ b. $\frac{1}{2}$ c. $\frac{1}{3}$ d. $\frac{1}{8}$ e. 1

Estimation 5 Addition and Subtraction of Fractions

1. $\frac{2}{3} + \frac{1}{3} =$ ______ 2. $\frac{2}{3} - \frac{1}{3} =$ ______ 3. $\frac{5}{8} - \frac{1}{4} =$ ______ 4. $\frac{5}{8} + \frac{1}{4} =$ ______

Answers: a. 1 b. $\frac{1}{2}$ c. $\frac{3}{4}$ d. $\frac{3}{2}$ e. $\frac{1}{4}$

Estimation 6 Addition and Subtraction of Fractions

1. $\frac{7}{8} - \frac{1}{2} =$ ______ 2. $\frac{2}{3} + \frac{1}{4} =$ ______ 3. $\frac{2}{3} + \frac{3}{4} =$ ______ 4. $\frac{3}{4} - \frac{1}{4} =$ ______

Answers: a. $\frac{3}{2}$ b. 2 c. 1 d. $\frac{5}{4}$ e. $\frac{1}{2}$

Estimation 7 Multiplication of Fractions

1. $\frac{1}{2} \times \frac{1}{4} =$ ______ 2. $\frac{1}{2} \times \frac{1}{3} =$ ______ 3. $\frac{1}{3} \times \frac{1}{6} =$ ______ 4. $\frac{1}{5} \times \frac{1}{4} =$ ______

Answers: a. $\frac{1}{10}$ b. $\frac{1}{2}$ c. $\frac{1}{8}$ d. $\frac{1}{6}$ e. $\frac{1}{20}$

Estimation 8 Multiplication of Fractions

1. $\frac{2}{3} \times \frac{3}{4} =$ ______ 2. $\frac{4}{5} \times \frac{1}{2} =$ ______ 3. $\frac{5}{6} \times \frac{1}{4} =$ ______ 4. $\frac{2}{5} \times \frac{3}{5} =$ ______

Answers: a. 1 b. $\frac{1}{4}$ c. $\frac{1}{2}$ d. $\frac{3}{4}$ e. $\frac{1}{3}$

Estimation 9 Division of Fractions

1. $\frac{1}{2} \div \frac{1}{4} =$ ______ 2. $\frac{1}{2} \div \frac{1}{3} =$ ______ 3. $\frac{1}{3} \div \frac{1}{4} =$ ______ 4. $\frac{1}{5} \div \frac{1}{4} =$ ______

Answers: a. 1 b. 2 c. $\frac{3}{2}$ d. $\frac{3}{4}$ e. $\frac{5}{4}$

Add and Subtract Fractions

Directions

Add these fractions.

a. $\frac{1}{2} + \frac{1}{3}$ ______

b. $\frac{3}{5} + \frac{2}{5}$ ______

c. $\frac{1}{2} + \frac{2}{3}$ ______

d. $\frac{3}{4} + \frac{2}{3}$ ______

e. $\frac{2}{5} + \frac{1}{3}$ ______

f. $\frac{7}{9} + \frac{5}{7}$ ______

Subtract these fractions.

a. $\frac{1}{2} - \frac{1}{3}$ ______

b. $\frac{3}{5} - \frac{2}{5}$ ______

c. $\frac{2}{3} - \frac{1}{2}$ ______

d. $\frac{3}{4} - \frac{2}{3}$ ______

e. $\frac{2}{5} - \frac{1}{3}$ ______

f. $\frac{7}{9} - \frac{5}{7}$ ______

Add these mixed numbers.

a. $1\frac{1}{2} + \frac{1}{4}$ ______

b. $5\frac{6}{7} + 2\frac{1}{8}$ ______

c. $\frac{7}{6} + 1\frac{5}{6}$ ______

d. $\frac{7}{3} + 6\frac{1}{7}$ ______

e. $1\frac{3}{4} + 2\frac{1}{8}$ ______

f. $5\frac{1}{9} + 2$ ______

Subtract these mixed numbers.

a. $3\frac{1}{4} - \frac{3}{2}$ ______

b. $5\frac{6}{7} - 2\frac{1}{8}$ ______

c. $\frac{11}{6} - \frac{7}{6}$ ______

d. $2\frac{1}{8} - 1\frac{3}{4}$ ______

e. $6\frac{1}{7} - \frac{7}{3}$ ______

f. $5\frac{1}{9} - 2$ ______

Multiply and Divide Fractions

Directions

Simplify your answers using the [Simp] and [=] keys.

Multiply these fractions.

a. $\frac{1}{2} \times \frac{1}{3}$ ________

b. $\frac{3}{5} \times \frac{2}{5}$ ________

c. $\frac{1}{2} \times \frac{2}{3}$ ________

d. $\frac{3}{4} \times \frac{2}{3}$ ________

e. $\frac{2}{5} \times \frac{1}{3}$ ________

f. $\frac{7}{9} \times \frac{5}{7}$ ________

Divide these fractions.

a. $\frac{1}{2} \div \frac{1}{3}$ ________

b. $\frac{3}{5} \div \frac{2}{5}$ ________

c. $\frac{1}{2} \div \frac{2}{3}$ ________

d. $\frac{3}{4} \div \frac{2}{3}$ ________

e. $\frac{2}{5} \div \frac{1}{3}$ ________

f. $\frac{7}{9} \div \frac{5}{7}$ ________

Multiply these mixed numbers.

a. $\frac{3}{2} \times 1\frac{1}{4}$ ________

b. $5\frac{6}{7} \times 2\frac{1}{8}$ ________

c. $3\frac{7}{6} \times \frac{11}{6}$ ________

d. $\frac{7}{3} \times 6\frac{1}{7}$ ________

e. $1\frac{3}{4} \times 2\frac{1}{8}$ ________

f. $5\frac{1}{9} \times 2$ ________

Divide these mixed numbers.

a. $\frac{3}{2} \div 1\frac{1}{4}$ ________

b. $5\frac{6}{7} \div 2\frac{1}{8}$ ________

c. $3\frac{7}{6} \div \frac{11}{6}$ ________

d. $\frac{7}{3} \div 6\frac{1}{7}$ ________

e. $1\frac{3}{4} \div 2\frac{1}{8}$ ________

f. $5\frac{1}{9} \div 2$ ________

Rule of Thumb

Directions

1. Perform the following computations on the Math Explorer. Write the answer to the last problem. Then write the problem for the answer. Look for a pattern.

   a. $\frac{1}{2} + \frac{1}{2} =$ ________ $\times \frac{2}{2} =$ ________

   b. $\frac{1}{2} + \frac{1}{3} =$ ________

   c. $\frac{1}{2} + \frac{1}{4} =$ ________ $\times \frac{2}{2} =$ ________

   d. $\frac{1}{2} + \frac{1}{5} =$ ________

   e. $\frac{1}{2} + \frac{1}{6} =$ ________ $\times \frac{2}{2} =$ ________

   f. ___ + ___ = ________

2. Write the next answer. Then write the correct problem for that answer and verify it on the Math Explorer.

3. Perform the following computations. Look for a pattern.

   a. $\frac{1}{2} + \frac{2}{3} =$ ________

   b. $\frac{1}{2} + \frac{3}{4} =$ ________ $\times \frac{2}{2} =$ ________

   c. $\frac{1}{2} + \frac{4}{5} =$ ________

   d. $\frac{1}{2} + \frac{5}{6} =$ ________ $\times \frac{2}{2} =$ ________

   e. $\frac{1}{2} + \frac{6}{7} =$ ________

   f. ___ + ___ = ________

Write the answer to the last problem. Then write the problem for the answer.

4. Write the next answer. Then write the correct problem for that answer and verify it on the Math Explorer.

5. Solve the following problems mentally. Check your answers on the Math Explorer. Try to develop a rule for the addition of fractions that will allow you to work the problems mentally.

a. $\frac{1}{4} + \frac{1}{5}$ d. $\frac{2}{5} + \frac{1}{4}$ g. $\frac{2}{5} + \frac{2}{7}$

b. $\frac{1}{4} + \frac{1}{9}$ e. $\frac{3}{4} + \frac{1}{5}$ h. $\frac{2}{7} + \frac{3}{8}$

c. $\frac{1}{5} + \frac{1}{6}$ f. $\frac{2}{5} + \frac{1}{3}$ i. $\frac{4}{9} + \frac{2}{7}$

Rule of Thumb, Jr.

Directions

1. Perform the following computations. Look for a pattern.

a. $\frac{2}{3} - \frac{1}{2} =$ ____

b. $\frac{3}{4} - \frac{1}{2} =$ ____ $\times \frac{2}{2}$ ____

c. $\frac{4}{5} - \frac{1}{2} =$ ____

d. $\frac{5}{6} - \frac{1}{2} =$ ____ $\times \frac{2}{2}$ ____

e. $\frac{2}{3} - \frac{1}{3} =$ ____ $\times \frac{3}{3}$ ____

f. $\frac{3}{4} - \frac{1}{3} =$ ____

g. $\frac{4}{5} - \frac{1}{3} =$ ____

h. $\frac{5}{6} - \frac{1}{3} =$ ____ $\times \frac{3}{3}$ ____

2. Write the answer to the next problem in each column. Now write the problems for the answers you just computed.

____________________ ____________________

3. Write the next answer. Now write the correct problem for the answer you just gave. Verify the problem using the Math Explorer.

____________________ ____________________

4. Solve the following problems mentally. Check your answers on the Math Explorer. Try to develop a rule for subtraction of fractions that will allow you to work them mentally.

a. $\frac{3}{4} - \frac{2}{5} =$ ____

b. $\frac{3}{4} - \frac{2}{3} =$ ____

c. $\frac{5}{3} - \frac{1}{5} =$ ____

d. $\frac{8}{9} - \frac{5}{6} =$ ____

e. $\frac{4}{5} - \frac{2}{3} =$ ____

f. $\frac{4}{5} - \frac{2}{7} =$ ____

g. $\frac{2}{7} - \frac{1}{6} =$ ____

h. $\frac{5}{9} - \frac{1}{4} =$ ____

Best Guess

Directions

Estimate and circle the answer you think is the best for each problem. Then use the Math Explorer to check your estimates.

1. Multiply 4759 by $\frac{1}{34}$. The best answer is:

   a. 150 b. 15.7 c. 1500 d. 1.57

2. Multiply 4759 by $\frac{34}{11}$. The best answer is:

   a. 150 b. 15.0 c. 15,000 d. 1500

3. Multiply 4759 by $\frac{9}{11}$. The best answer is:

   a. 400 b. 40.8 c. 40,000 d. 4000

4. Multiply $\frac{11}{32}$ by $\frac{1}{5}$. The best answer is:

   a. $\frac{55}{32}$ b. 160 c. $\frac{11}{160}$ c. 1760

5. Multiply $\frac{1}{16}$ by $\frac{9}{5}$. The best answer is:

   a. $\frac{9}{80}$ b. 720 c. $\frac{10}{21}$ c. $\frac{5}{144}$

6. Divide 4759 by $\frac{1}{14}$. The best answer is:

   a. 6700 b. 67,000 c. 670 d. 67

7. Divide 4759 by $\frac{5}{8}$. The best answer is:

   a. 77,000 b. 770 c. 7700 d. 77.7

8. Divide 4759 by $\frac{23}{5}$. The best answer is:

   a. 1100 b. 11,000 c. 110 d. 1.10

9. Divide $\frac{11}{32}$ by $\frac{1}{5}$. The best answer is:

   a. $\frac{55}{32}$ b. $\frac{11}{160}$ c.1760 d. 160

10. Divide $\frac{1}{16}$ by $\frac{9}{5}$. The best answer is:

    a. $\frac{9}{80}$ b. 720 c. $\frac{10}{21}$ d. $\frac{5}{144}$

Good, Better, Best

Directions

Estimate and circle the answer you think is the best for each problem. Then use the Math Explorer to check your estimates.

1. Add $\frac{7}{16}$ to $\frac{5}{9}$. The best answer is:

   a. $\frac{12}{25}$ b. $\frac{143}{144}$ c. 37 d. $\frac{63}{80}$

2. Add $\frac{12}{5}$ to $\frac{1}{24}$. The best answer is:

   a. $\frac{293}{120}$ b. $\frac{12}{120}$ c. $\frac{13}{29}$ d. 41

3. Add $\frac{29}{7}$ to $\frac{2}{3}$. The best answer is:

   a. $\frac{58}{21}$ b. $\frac{21}{58}$ c. 41 d. $\frac{101}{21}$

4. Add $\frac{5}{12}$ to 24. The best answer is:

   a. $\frac{29}{12}$ b. $\frac{293}{12}$ c. $\frac{5}{36}$ d. 39

5. Subtract $\frac{5}{12}$ from 24. The best answer is:

   a. $\frac{283}{12}$ b. $\frac{19}{24}$ c. $\frac{31}{36}$ d. 31

6. Subtract $\frac{2}{3}$ from $\frac{29}{7}$. The best answer is:

   a. $\frac{27}{4}$ b. $\frac{73}{21}$ c. $\frac{27}{21}$ d. 21

7. Subtract $\frac{1}{15}$ from $\frac{19}{24}$. The best answer is:

   a. $\frac{-87}{120}$ b. $\frac{87}{120}$ c. $\frac{18}{39}$ d. 2

8. Subtract $\frac{7}{8}$ from $\frac{10}{11}$. The best answer is:

   a. $\frac{3}{3}$ b. $\frac{3}{88}$ c. $\frac{-3}{88}$ d. $\frac{3}{19}$

Tables of Stone?

Directions

1. Given that [ ] + { } = $\frac{7}{11}$ and # - @ = $\frac{2}{9}$, complete the table.

	[ ]	{ }	#	@
a.	$\frac{1}{2}$	____	$\frac{5}{6}$	____
b.	$\frac{2}{3}$	____	$\frac{9}{19}$	____
c.	$\frac{4}{7}$	____	$\frac{4}{7}$	____
d.	____	$\frac{3}{8}$	____	$\frac{1}{8}$
e.	____	$\frac{3}{13}$	____	$\frac{2}{13}$

2. Given that $y = \frac{a}{b} + \frac{1}{2}$ and $n = \frac{1}{3} + \frac{x}{y}$, complete the table.

	$\frac{a}{b}$	y	$\frac{x}{y}$	n
a.	$\frac{1}{4}$	$\frac{3}{4}$	$\frac{1}{6}$	$\frac{1}{2}$
b.	$\frac{1}{6}$	____	$\frac{2}{9}$	____
c.	$\frac{3}{10}$	____	____	$\frac{5}{6}$
d.	____	$\frac{9}{14}$	____	$\frac{13}{21}$
e.	____	$\frac{9}{10}$	$\frac{13}{33}$	____

Lesson 13: Converting Improper Fractions to Mixed Numbers

Overview

This lesson shows students how to use the Math Explorer to change improper fractions to mixed numbers and how to switch back and forth between the improper fraction and its mixed number equivalent.

Transparencies

Transparency 19, *Converting Improper Fractions*, Chapter 2

Keys Introduced

[Ab/c] [1/x]

Student Worksheets

Checking Squares
Ordered Drill
Ups and Downs
A Mixed Bag
Improper Changes

Teaching Steps

1. Use Transparency 19, *Converting Improper Fractions*, to demonstrate how to convert improper fractions to mixed numbers.
2. Show that, besides being entered from the keyboard, improper fractions can also result from a calculation.

Press	Display
2 [/] 3 [+]	+ 2/3
1 [/] 2 [=]	7/6

3. Ask students, "What do you notice about the denominator of $\frac{7}{6}$ compared with $\frac{2}{3}$ and $\frac{1}{2}$?" (It's the lowest common denominator.)

Five-Minute Filler

To change temperatures from degrees Fahrenheit to degrees Celsius, the formula $C = \frac{5}{9}(F-32)$ is used. One day in May it was 107 degrees Fahrenheit in Phoenix, Arizona, and 3 degrees Fahrenheit in Barrow, Alaska. What is the difference in degrees Celsius between the two temperatures?

Checking Squares

Directions

1. Record the whole-number portion of each mixed number in the same position in the adjacent table. Check to see whether the resulting square is a magic square. Use the [Ab/c] key.

$\frac{51}{8}$	$\frac{93}{89}$	$\frac{75}{9}$
$\frac{37}{5}$	$\frac{77}{15}$	$\frac{137}{45}$
$\frac{15}{7}$	$\frac{78}{8}$	$\frac{259}{64}$

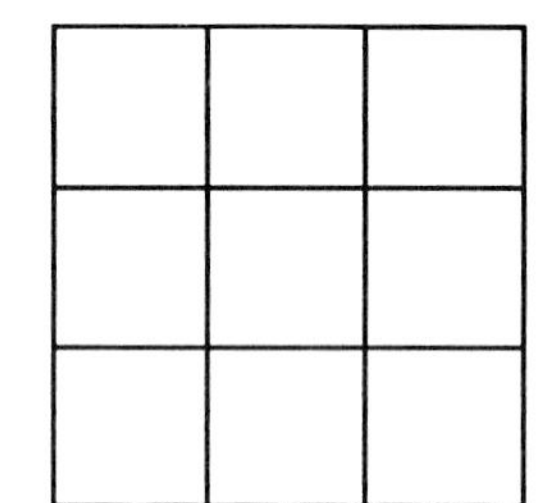

______

2. Record the numerator of the fractional part of each mixed number in the same positon in the adjacent table. Check to see whether the resulting square is a magic square.

$\frac{48}{7}$	$\frac{31}{5}$	$\frac{17}{9}$
$\frac{77}{8}$	$\frac{131}{18}$	$\frac{118}{23}$
$\frac{16}{7}$	$\frac{100}{13}$	$\frac{157}{17}$

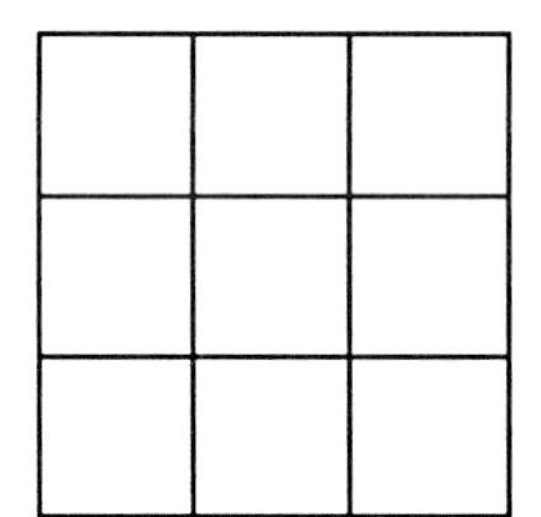

______

3. Add the whole-number part and the numerator of the fractional part of the mixed number together and record the sum in the same positon in the adjacent table. Check to see whether the resulting square is a magic square.

$\frac{90}{7}$	$\frac{11}{5}$	$\frac{168}{7}$
$\frac{61}{3}$	$\frac{155}{11}$	$\frac{569}{71}$
$\frac{54}{13}$	$\frac{227}{9}$	$\frac{242}{24}$

______

Ordered Drill

Directions

1. Write a symbol for "greater than" or "less than," > or <, on each blank by guessing.
2. Convert the improper fractions to mixed numbers using the [Ab/c] key to decide whether you have inserted the inequality properly.

a. $\frac{45}{7}$ ______ $\frac{32}{5}$

b. $\frac{12}{5}$ ______ $\frac{8}{3}$

c. $\frac{63}{5}$ ______ $\frac{35}{2}$

d. $\frac{77}{10}$ ______ $\frac{48}{7}$

e. $\frac{94}{7}$ ______ $\frac{87}{6}$

f. $\frac{8}{5}$ ______ $\frac{19}{12}$

g. $\frac{66}{8}$ ______ $\frac{46}{4}$

h. $\frac{60}{16}$ ______ $\frac{75}{32}$

i. $\frac{125}{64}$ ______ $\frac{80}{32}$

j. $\frac{32}{12}$ ______ $\frac{42}{9}$

k. $\frac{26}{8}$ ______ $\frac{30}{9}$

l. $\frac{13}{12}$ ______ $\frac{25}{23}$

Ups and Downs

Directions

1. Arrange the following fractions in descending value by guessing. Check your arrangement on the Math Explorer:

   a. $\frac{23}{3}$ b. $\frac{16}{3}$ c. $\frac{47}{13}$ d. $\frac{9}{2}$ e. $\frac{42}{8}$ f. $\frac{15}{4}$

   ____

   a. Enter the improper fraction into the display of the Math Explorer.

   b. Press the Ab/c key to change to a mixed number.

   c. Consider the whole numbers as you arrange the fractions order. If duplicate whole numbers appear, compare the fractional part of the mixed number.

2. Arrange the following fractions in ascending value by guessing. Check your arrangement on the Math Explorer.

   a. $\frac{51}{4}$ b. $\frac{49}{9}$ c. $\frac{18}{5}$ d. $\frac{32}{3}$ e. $\frac{57}{39}$ f. $\frac{63}{23}$

   ____

A Mixed Bag

Directions

1. Enter each mixed number into the Math Explorer using the [Unit] key after you enter the whole number.
2. To change the entered mixed number to an improper fraction, press the [1/x] key twice.

a. $2\frac{3}{4}$	e. $7\frac{5}{8}$	i. $83\frac{1}{3}$	m. $37\frac{6}{7}$
b. $5\frac{1}{3}$	f. $8\frac{3}{7}$	j. $76\frac{4}{5}$	n. $3\frac{75}{76}$
c. $6\frac{2}{7}$	g. $12\frac{5}{9}$	k. $46\frac{7}{10}$	o. $40\frac{8}{12}$
d. $4\frac{3}{5}$	h. $13\frac{5}{9}$	l. $5\frac{13}{14}$	p. $3\frac{8}{90}$

Improper Changes

Directions

Fill in the blanks with either a mixed number or an improper fraction.

Mixed number	Improper fraction	Mixed number	Improper fraction
$5\frac{3}{4}$	______	______	$\frac{56}{7}$
______	$\frac{89}{64}$	$16\frac{13}{14}$	______
$4\frac{5}{6}$	______	$17\frac{5}{23}$	______
______	$\frac{121}{17}$	______	$\frac{111}{9}$
______	$\frac{50}{9}$	$22\frac{15}{39}$	______
$17\frac{6}{17}$	______	______	$\frac{86}{7}$
$10\frac{1}{10}$	______	$4\frac{4}{11}$	______
______	$\frac{19}{6}$	$36\frac{9}{29}$	______
______	$\frac{101}{10}$	______	$\frac{52}{7}$

Lesson 14: Mixed-Number Operations

Overview

This lesson shows students how to use the Math Explorer to perform addition, subtraction, multiplication, and division of fractions.

Transparencies

Transparency 20, *Math Operations: Fractions*, Chapter 2

Keys Introduced

None

Student Worksheets

Another Mixed Bag
The Trouble Maker
Is It Really Magic?
The Ancients of Noitcarf

Teaching Steps

1. Use Transparency 20, *Math Operations: Fractions* to demonstrate fractions operations.
2. Review the use of the [Unit] key and its role in entering mixed numbers into the calculator display.

Press	**Display**
2 [Unit] 3 [/] 8	2 u 3/8

3. Have the students complete *Another Mixed Bag* worksheet.

Five-Minute Filler

A blueprint has been reduced to $8\frac{1}{2}$ inches by $10\frac{3}{4}$ inches. This is one-third its original size. What were its original dimensions and area?

Another Mixed Bag

Directions

Perform the indicated operation. Be sure the answer is expressed in simplest terms.

a. $1\frac{1}{2} + \frac{1}{4}$ ______

b. $\frac{7}{6} + 1\frac{5}{6}$ ______

c. $2\frac{1}{8} - 1\frac{3}{4}$ ______

d. $1\frac{1}{5} \times 2\frac{2}{7}$ ______

e. $6\frac{1}{4} \times 10$ ______

f. $12 - 4\frac{1}{4}$ ______

g. $\frac{5}{6} \div 1\frac{3}{5}$ ______

h. $3\frac{1}{2} \div 2\frac{1}{3}$ ______

i. $3\frac{2}{3} \times 1\frac{1}{4}$ ______

j. $3\frac{1}{3} + 9$ ______

k. $2\frac{1}{4} + 1\frac{2}{3}$ ______

l. $4\frac{1}{5} - 1\frac{2}{7}$ ______

m. $7\frac{1}{2} \div \frac{7}{10}$ ______

n. $4\frac{1}{3} \div \frac{3}{7}$ ______

o. $1\frac{2}{5} - \frac{5}{8}$ ______

Peter is making some plates for a conveyer system. Each plate must be $3\frac{3}{4}$ inches long. He has a metal strip $56\frac{1}{2}$ inches long from which he is to make the plates. How many plates will he be able to cut from the strip?

______

The Trouble Maker

Directions

1. Perform the indicated operation without using the Math Explorer if you can.
2. Be sure the answer is written as a mixed number in simplest terms.
3. Enter the numerator of the fraction part of each answer in the Math Explorer.
4. Turn the Math Explorer upside to read each word. Put the words together to discover what Rob hid that got him fired from his job.

a. $2\frac{1}{2} + 6\frac{1}{3} =$ ____

b. $3\frac{1}{4} \times \frac{1}{13} =$ ____

c. $1\frac{16}{20} - 1 =$ ____

____ ____ ____ ____

d. $2\frac{1}{2} \div 3\frac{1}{2} =$ ____

e. $6\frac{1}{2} \div 5\frac{1}{4} =$ ____

f. $7\frac{1}{6} - 2\frac{3}{4} =$ ____

g. $2\frac{5}{6} \div \frac{17}{24} =$ ____

h. $\frac{2}{3} \times \frac{4}{5} =$ ____

____ ____ ____ ____

i. $4\frac{7}{8} \times 3 =$ ____

j. $7 \div 4\frac{9}{10} =$ ____

k. $9\frac{5}{7} - 3\frac{15}{21} =$ ____

l. $14\frac{3}{4} \times 16\frac{4}{5} =$ ____

m. $2\frac{1}{2} \times 18\frac{1}{3} =$ ____

____ ____ ____ ____ ____ ____

Is It Really Magic?

Directions

1. A magic square is an array of numbers in which the sum of numbers in each row, column, and diagonal is the same. This sum is called the magic number.
2. Use the given numbers to find the magic number for each square. Then tell whether each square can be completed to make it a magic square. The rows, columns and diagonals all have the same sum.

		$\frac{2}{3}$
	$\frac{5}{6}$	$1\frac{1}{2}$
1		

______

$2\frac{1}{4}$	$\frac{3}{8}$	
$2\frac{5}{8}$		$1\frac{1}{8}$
$\frac{3}{4}$		

______

$1\frac{1}{4}$	$2\frac{7}{8}$	$\frac{3}{4}$
	$\frac{5}{6}$	
		$5\frac{19}{24}$

______

5		
	$4\frac{1}{6}$	
$1\frac{2}{3}$		$3\frac{1}{3}$

______

The Ancients of Noitcarf

The Ancients of Noitcarf ("fraction" spelled backward) is a rule-finding game involving fractions. You must find the rule in order to complete the pyramid and find the key.

The Situation

While exploring the ruins at the city of Noitcarf, you come upon a chamber. You know that a treasure and many wonderful secrets are inside. You find three keys with writing on them, which proves they were made by the Ancients of Noitcarf! Only one of the keys will open the chamber. If you use the wrong key, the chamber will vanish forever. You notice a mysterious drawing on the wall. You realize that the top block on the drawing will tell which key to choose.

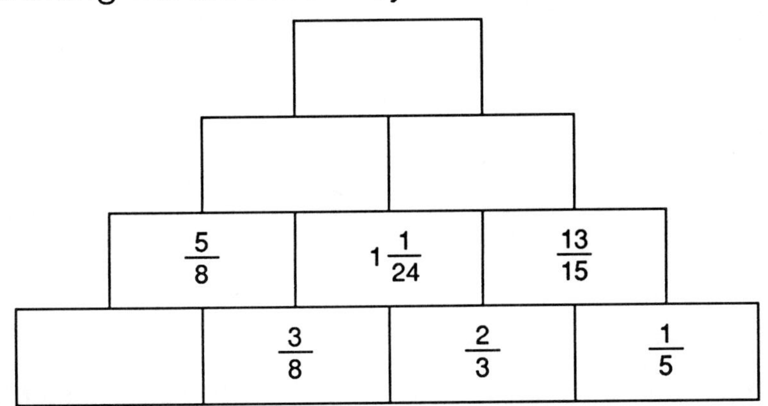

Directions

1. Find the rule that the Ancients of Noitcarf used in the pyramid.
2. Once you figure out the rule, complete the pyramid.
3. The answer in the top block will tell you which key to choose. Circle the correct key.

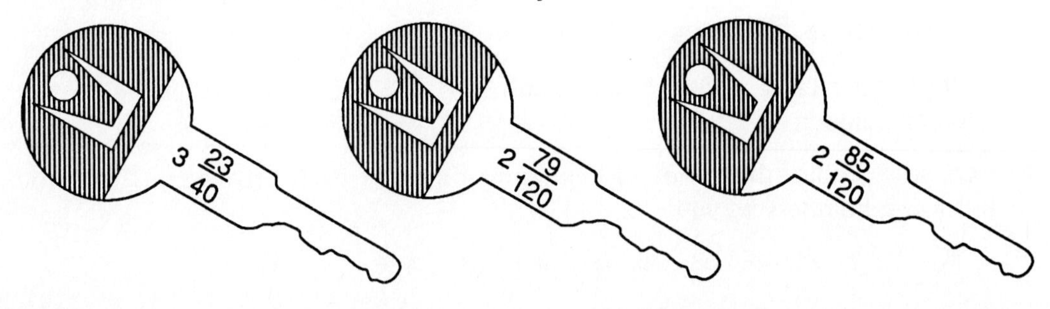

Lesson 15: Operations on Decimals

Overview

Students learn to perform basic operations involving decimals on the calculator.

Transparencies

Transparency 16, *Fixing the Decimal Point*, Chapter 2

Keys Introduced

None

Student Worksheets

A Record Event
Determination
What Does That Point Prove?
Redundancy
Step Along
Four Straight

Teaching Steps

1. Use Transparency 16, *Fixing the Decimal Point*, to review that function on the calculator. Also review how to determine the appropriate number of decimal places in an answer.
2. Review the need to estimate answers. Point out that with decimals it is easy to have digits right but the order of magnitude wrong because of incorrectly placed decimal points when entering numbers.
3. Have the students estimate and record answers before they are allowed to work the problems on the worksheets in this lesson on the Math Explorer.

Five-Minute Fillers

1. Multiply your favorite single-digit number by 0.9 and then by 12,345,679. Try another number.
2. Can you predict the series of repeating digits in each answer? Divide the following numbers by 999:

   4, 6, 9, 12, 25, 46, 63, 76, 89, 462, 238, 454

A Record Event

Directions

1. Record the number of decimal places in each factor in the space by the factor.
2. Estimate the answer.
3. Use the Math Explorer to get the exact answer.

a. 3.9 ____
× 5.2 ____

______ Estimate

______ Actual

b. 13.29 ____
× 15.36 ____

______ Estimate

______ Actual

c. 8.31 ____
× 2.9 ____

______ Estimate

______ Actual

d. 0.043 ____
× 7.9 ____

______ Estimate

______ Actual

e. 4.03 ____
× 12 ____

______ Estimate

______ Actual

f. 0.123 ____
× 0.89 ____

______ Estimate

______ Actual

g. 432.9 ____
× 0.586 ____

______ Estimate

______ Actual

h. 999.9 ____
× 0.998 ____

______ Estimate

______ Actual

i. 0.054 ____
× 0.54 ____

______ Estimate

______ Actual

4. Write the rule for locating the decimal point in multiplication problems.

____

____

Determination

Directions

1. Determine the number of decimal places in each answer.
2. Enter the number of decimal places in the square with the letter that corresponds to the problem.
3. Determine whether a magic square results.
4. Work the problems on the Math Explorer to see whether the Math Explorer gets the same number of decimal places as you entered in the square.

a. 0.06×0.78

b. 5.5×45

c. 0.144×0.12

d. 0.25×3.9

e. 4.567×1.207

f. 0.00079×0.0045

g. $8.57 \times 4{,}568$

h. 0.0029×0.9085

i. 0.0000078×24

e.	b.	h.
i.	c.	d.
g.	f.	a.

Is this a magic square? ______

What Does That Point Prove?

Directions

1. Write > or < on each blank.

   a. 0.04 __ 0.008

   b. 0.9 __ 0.99

   c. 0.328 __ 0.33

   d. 0.0792 __ 0.11

   e. 0.0057 __ 0.006

   f. 0.4 __ 0.0444

2. Arrange each of the following sets of numbers from smallest to largest.

   a. 0.03, 0.33, 0.033, 0.303 ____________________

   b. 0.017, 0.2, 0.02, 0.007 ____________________

   c. 0.4, 0.405, 0.45, 0.045 ____________________

   d. 0.04, 0.304, 0.32, 0.4 ____________________

   e. 0.2, 0.06, 0.0602, 0.026 ____________________

   f. 0.082, 0.28, 0.8, 0.08 ____________________

3. Check your answers with the Math Explorer by entering the decimal into the display and pressing the [1/x]. The decimal with the smallest reciprocal value is the largest decimal.

Redundancy

Directions

1. Work the first problem of each set on the Math Explorer. Then estimate the answer for each of the other problems in the set.

   Check your answers on the Math Explorer.

   a. 0.0056 ÷ 0.0007 ______
   b. 5.6 ÷ 0.7 ______
   c. 0.056 ÷ 0.007 ______
   d. 56 ÷ 7 ______
   e. 0.56 ÷ 0.07 ______

   f. 0.01020 ÷ 0.03 ______
   g. 0.1020 ÷ 0.3 ______
   h. 1.020 ÷ 3 ______
   i. 10.2 ÷ 30 ______
   j. 102.0 ÷ 300 ______
   k. 1020 ÷ 3000 ______

2. Tell why all the answers in each set are the same.

   ____

   ____

3. Make up a division exercise for each set of answers.

   a. 0.00108 ÷ 0.036 = 0.03
   b. ______ ÷ ______ = 0.03
   c. ______ ÷ ______ = 0.03
   d. ______ ÷ ______ = 0.03
   e. ______ ÷ ______ = 0.03

   f. 0.01804 ÷ 8.2 = 0.0022
   g. ______ ÷ ______ = 0.0022
   h. ______ ÷ ______ = 0.0022
   i. ______ ÷ ______ = 0.0022
   j. ______ ÷ ______ = 0.0022

Step Along

Directions

Record the answer to each step as you solve the problems using the Math Explorer.

a. $72 \div 0.009 \times 0.125 \div 1000$ ____ ____ ____

b. $84 \div 0.48 \times 36 \div 0.63 \times 0.0001$ ____ ____ ____ ____

c. $0.54 \div 0.06 \times 0.05 \div 0.45$ ____ ____ ____

d. $(14 + 35) \div 0.14 \div 0.35 \times 0.001$ ____ ____ ____ ____

e. $120 \times 0.01 \div 0.06 \times 0.05$ ____ ____ ____

f. $0.63 \div 0.36 \times 0.12 \div 0.21$ ____ ____ ____

g. $(12 + 60) \div 0.12 \div 0.6 \times 0.001$ ____ ____ ____ ____

h. $0.42 \div 0.24 \times 12 \div 21$ ____ ____ ____

Four Straight

Directions

1. Take a good look at the answers in the box below.
2. Use estimation to select four problems that will give four answers in a row in the box. (These answers can occur horizontally, vertically, or diagonally.)
3. Use the Math Explorer to determine whether your problems give four answers in a row.

   a. 200.7 + 20.07 + 2.007 + 0.2007

   b. 0.595 + 0.68 + 36.49

   c. 351.8 + 86.42 + 7.268 + 0.8706

   d. 4.667 + 0.312 + 15.08

   e. 817.4 + 6.2764 + 0.768 + 89.6

   f. 3.005 + 12.00724 + 87.9

   g. 63 + 5.1 + 132.172 + 18.83

   h. 80.2 + 0.0312 + 267.3701

   i. 0.9876 + 9.876 + 98.76 + 987.6

   j. 0.1644 + 796.1033 + 10.616

1.2754	347.6013	446.3586	914.0444
102.91224	796.278	37.765	53.2451
45.678	222.9777	20.059	806.8837
219.102	345.786	1,097.2236	78.93456

Lesson 16: Changing Between Fractions and Decimals

Overview

Students learn to change fractions to decimals and decimals to fractions.

Transparencies

Transparency 22, *Fractions and Decimals*, Chapter 2

Keys Introduced

None

Student Worksheets

Decimal Fractions
Completion Test
Order Up!
Converted
Equivalent Values?
Find the Fraction

Teaching Steps

1. Use Transparency 22, *Fractions and Decimals*, to demonstrate the relationship between fractions and decimals.
2. Point out that decimals are another form in which fractions can be written and operated upon.

Five-Minute Filler

Express as a decimal the time students watch each type of movie.

Comedy $\frac{1}{2}$ Drama $\frac{1}{4}$

History $\frac{1}{12}$ Mystery $\frac{1}{6}$

Decimal Fractions

Directions

1. Enter each fraction into the Math Explorer.
2. Press the [F↔D] key. Record the decimal shown in the display of the Math Explorer.
3. Try to change the decimal back into a fraction by pressing the [F↔D] key and then simplifying any resulting fraction.
4. Circle the fractions that can be changed to decimals and then back to the original fractions by the Math Explorer when the above procedure has been followed.

a. $\frac{3}{4}$ ______ g. $\frac{7}{20}$ ______

b. $\frac{2}{3}$ ______ h. $\frac{1}{12}$ ______

c. $\frac{5}{8}$ ______ i. $\frac{2}{7}$ ______

d. $\frac{2}{5}$ ______ j. $\frac{4}{5}$ ______

e. $\frac{1}{6}$ ______ k. $\frac{5}{18}$ ______

f. $\frac{7}{11}$ ______ l. $\frac{5}{9}$ ______

Completion Test

Directions

1. Find the decimal equivalent of the first four fractions in each set using the Math Explorer.
2. Complete the set using the pattern that has developed.
3. Check your answers using the Math Explorer.

a. $\frac{1}{9}$ = ________

$\frac{2}{9}$ = ________

$\frac{3}{9}$ = ________

$\frac{4}{9}$ = ________

$\frac{5}{9}$ = ________

$\frac{6}{9}$ = ________

c. $\frac{1}{7}$ = ________

$\frac{2}{7}$ = ________

$\frac{3}{7}$ = ________

$\frac{4}{7}$ = ________

$\frac{5}{7}$ = ________

$\frac{6}{7}$ = ________

b. $\frac{1}{11}$ = ________

$\frac{2}{11}$ = ________

$\frac{3}{11}$ = ________

$\frac{4}{11}$ = ________

$\frac{5}{11}$ = ________

$\frac{6}{11}$ = ________

$\frac{7}{11}$ = ________

$\frac{8}{11}$ = ________

$\frac{9}{11}$ = ________

$\frac{10}{11}$ = ________

d. $\frac{1}{12}$ = ________

$\frac{2}{12}$ = ________

$\frac{3}{12}$ = ________

$\frac{4}{12}$ = ________

$\frac{5}{12}$ = ________

$\frac{6}{12}$ = ________

$\frac{7}{12}$ = ________

$\frac{8}{12}$ = ________

$\frac{9}{12}$ = ________

$\frac{10}{12}$ = ________

4. Why is the last digit often one more than you would expect from the previous outcomes? ______________________________

____

Order up!

Directions

1. Arrange the following fractions from least to greatest by guessing.

   a. $\frac{7}{32}$ b. $\frac{3}{16}$ c. $\frac{3}{5}$ d. $\frac{5}{18}$ e. $\frac{1}{2}$ f. $\frac{1}{49}$ g. $\frac{6}{19}$

   ____ ____ ____ ____ ____ ____ ____

2. Change the fractions to decimals by entering them into the Math Explorer. Press the [F↔D] key.

   ____ ____ ____ ____ ____ ____ ____

3. Arrange the decimal values from smallest to largest and compare the order with your guess.

   ____ ____ ____ ____ ____ ____ ____

4. Arrange the following fractions from greatest to least by guessing.

   a. $\frac{2}{21}$ b. $\frac{1}{8}$ c. $\frac{2}{32}$ d. $\frac{4}{17}$ e. $\frac{2}{3}$ f. $\frac{1}{9}$ g. $\frac{7}{20}$

   ____ ____ ____ ____ ____ ____ ____

5. Change the fractions to decimals to check your answers.

   ____ ____ ____ ____ ____ ____ ____

Converted

Directions

1. Use the [F↔D] key to convert the following decimals to fractions. Then use the [Simp] key followed by the [=] key to reduce the fractions to simplest terms.

a. 0.3 ______	f. 0.19 ______	k. 0.64 ______
b. 0.45 ______	g. 0.341 ______	l. 0.72 ______
c. 0.689 ______	h. 134.2 ______	m. 2.25 ______
d. 1.35 ______	i. 0.09 ______	n. 7.75 ______
e. 12.12 ______	j. 3.715 ______	o. 12.8 ______

2. Use the [F↔D] key to convert the following fractions to decimals. Then use the [Fix] key to round the answers to four decimal places.

a. $\frac{6}{25}$ ______	f. $\frac{131}{63}$ ______	k. $\frac{23}{19}$ ______
b. $\frac{5}{7}$ ______	g. $\frac{7}{12}$ ______	l. $\frac{5}{3}$ ______
c. $\frac{5}{9}$ ______	h. $\frac{3}{8}$ ______	m. $\frac{7}{32}$ ______
d. $\frac{5}{12}$ ______	i. $\frac{3}{7}$ ______	n. $\frac{31}{29}$ ______
e. $\frac{8}{11}$ ______	j. $\frac{11}{5}$ ______	0. $\frac{3}{13}$ ______

3. Arrange the following from least to greatest. Use the Math Explorer to check your listing.

a. 1.01 b. $\frac{7}{8}$ c. 0.048 d. $\frac{3}{8}$ e. 0.83

____ ____ ____ ____ ____

4. Arrange the following from greatest to least value. Use the Math Explorer to check your listing.

a. 0.40 b.0.375 c. $\frac{5}{12}$ d. $\frac{7}{10}$ e. $\frac{5}{6}$

____ ____ ____ ____ ____

Equivalent Values?

Directions

Construct on heavy paper the 32 game pieces shown.

$\frac{1}{4}$ 0.6	$\frac{5}{8}$ 0.6	$\frac{3}{4}$ 0.4	$\frac{5}{8}$ 0.4
$\frac{1}{3}$ 0.875	$\frac{9}{16}$ 0.875	$\frac{2}{8}$ 0.666	$\frac{3}{6}$ 0.666
$\frac{4}{16}$ 0.666	$\frac{3}{6}$ 0.666	$\frac{9}{12}$ 0.333	$\frac{10}{16}$ 0.333
$\frac{4}{6}$ 0.625	$\frac{10}{16}$ 0.625	$\frac{1}{4}$ 0.5625	$\frac{9}{12}$ 0.5625
$\frac{2}{8}$ 0.25	$\frac{3}{4}$ 0.25	$\frac{1}{3}$ 0.375	$\frac{8}{12}$ 0.375
$\frac{4}{16}$ 0.5	$\frac{9}{16}$ 0.625	$\frac{7}{8}$ 0.625	$\frac{7}{8}$ 075
$\frac{4}{6}$ 0.5	$\frac{9}{16}$ 0.925	$\frac{7}{8}$ 0.625	$\frac{7}{8}$ 0.75
$\frac{4}{10}$ 0.5	$\frac{4}{10}$ 0.25	$\frac{3}{5}$ 0.25	$\frac{3}{5}$ 0.25

Game Directions

1. Turn all game pieces face down.
2. Each player selects five pieces.
3. Choose which player will start. That player then selects one more piece and places it into the center of the playing area.
4. Player number 1 uses the Math Explorer to find the decimal equivalent for the fraction on the piece in the playing area.

   If player number 1 has the correct value on any of his or her pieces, that piece is placed end to end with its corresponding fraction. If player number 1 does not have the correct value in his or her pieces, he or she draws one piece at a time from the pile until a piece with the correct value is drawn or three pieces are drawn, whichever occurs first. If player number 1 draws three pieces without a match, it is player number 2's turn.

5. Player number 2 follows the same procedure as player number 1.
6. The winner is the player who uses all his or her pieces first.

Find the Fraction

Directions

Terminating decimals can be written as fractions whose denominators are powers of 10 (10, 100, 1000,...).

Infinite repeating decimals can be converted to fractions by following a sequence of steps similar to the following:

1. Represent the decimal by n: $n = 0.9090909090...$

2. Multiply both sides of the equation by a power of 10 that will move the decimal point past the first repeating pattern:

   $100n = 90.90909090....$

3. Subtract the equation in 1 from the equation in 2. This eliminates the infinite repeating part of the decimal.

$$\begin{aligned} 100n &= 90.90909090... \\ -n &= \underline{.90909090}... \\ 99n &= 90 \end{aligned}$$

4. Solve for n and reduce the fraction.

$$\begin{aligned} 99n &= 90 \\ n &= \frac{90}{99} \\ n &= \frac{1}{11} \end{aligned}$$

5. Find the fractional equivalent of the following decimal fractions.

   a. 0.3333... ________ d. 0.123123123...________

   b. 0.5555555.________ e. 0.222222... ________

   c. 0.656565...________ f. 0.272727... ________

Lesson 17: Changing Percents to Fractions and Decimals

Overview

Students learn to change fractions or decimals to equivalent percents, and percents to equivalent fractions and decimals.

Transparencies

Transparency 21, *Percents to Fractions*, Chapter 2

Keys Introduced

[%]

Student Worksheets

New Names
100 Per
Learn the Symbol
Percent Assigned
Mixed Sequence

Teaching Steps

1. Use Transparency 21, *Percents to Fractions*, to demonstrate the relationship between percents and fractions.
2. Point out that "percent" means "hundredths." In calculations, 14% is always written or used by the Math Explorer as 14/100 or 0.14.
3. Illustrate that a percent can be changed to a fraction in one or two easy steps. To change a percent to a fraction, write the percent as a fraction with 100 as the denominator. (It may be appropriate to reduce the fraction you obtain as an answer.)

Five-Minute Filler

If $\frac{1}{4}$ equals 25%, write equilvalent percents for the following fractions:

a. $\frac{2}{4}$ b. $\frac{3}{4}$ c. $\frac{4}{4}$ d. $\frac{5}{4}$ e. $\frac{6}{4}$

New Names

Directions

1. Change the following percents to fractions, as shown in the examples.
2. Check your work by entering the number into the display of the Math Explorer, pressing the [%] key, pressing the [F↔D] key, and finally reducing the fraction.

Examples: $35\% = 0.35 = \frac{35}{100} = \frac{7}{20}$

$5.4\% = 0.054 = \frac{54}{1000} = \frac{27}{500}$

a. 20% ________ ________ ________

b. 90% ________ ________ ________

c. 120% ________ ________ ________

d. 37.5% ________ ________ ________

e. 12.5% ________ ________ ________

f. 240% ________ ________ ________

100 Per

A percent can be changed to a decimal in two easy steps. First, write the percent as a fraction with a denominator of 100. (Can you see any similarity between the percent symbol and the numeral 100?) Then, the fraction (with a denominator of 100) can be written as a decimal by moving the decimal point in the numerator two places to the left.

Directions

1. Change the following percents to decimals as shown in the examples.
2. Check your work by entering the number into the display of the Math Explorer and then pressing the [%] key.

Examples: $56\% = \frac{56}{100} = 0.56$ $6.4\% = \frac{6.4}{100} = 0.064$

a. 36% ______________

b. 8.5% ______________

c. 9.2% ______________

d. 79% ______________

e. 13.41% ______________

f. 151% ______________

Learn the Symbol

A decimal can be changed to a percent by multiplying the decimal by 100 and writing a percent symbol (%).

Directions

1. Change each of the following decimals to an equivalent percent by entering the decimal in the Math Explorer.

2. Multiply by 100 and write a % symbol.

   Example: $0.625 \times 100 = 62.5\%$

a. 0.025 ________	g. 1.94 ________
b. 0.75 ________	h. 0.5 ________
c. 0.166 ________	i. 0.004 ________
d. 0.64 ________	j. 2.6 ________
e. 4.06 ________	k. 2.40 ________
f. 0.2 ________	l. 0.999 ________

3. Write a rule for changing any decimal to an equivalent percent without using the aid of the Math Explorer.

   ____

   ____

Percent Assigned

A fraction can be changed to a percent by dividing the numerator by the denominator, multiplying by 100, and writing a percent symbol (%).

Directions

1. Change each of the fractions listed to equivalent percents by entering the fraction in the Math Explorer.
2. Press the [F↔D] key.
3. Multiply by 100, round as appropriate, and write a % symbol.

Example: $1 \div 6 = 0.1666667 = 16.666667\% = 16.7\%$

a. $\frac{1}{2}$ ______

b. $\frac{2}{5}$ ______

c. $\frac{5}{8}$ ______

d. $\frac{1}{3}$ ______

e. $\frac{2}{9}$ ______

f. $1\frac{3}{8}$ ______

g. $\frac{6}{25}$ ______

h. $2\frac{3}{4}$ ______

i. $1\frac{1}{3}$ ______

j. $\frac{4}{7}$ ______

k. $3\frac{5}{12}$ ______

l. $5\frac{1}{20}$ ______

Mixed Sequence

Directions

1. Arrange the following collection of fractions, decimal fractions, and percentages in ascending order. First guess; then check using the Math Explorer.

	Guessed Sequence	Calculator Sequence
$\frac{2}{3}$	______	______
55%	______	______
0.80	______	______
$\frac{3}{4}$	______	______
33%	______	______
0.37	______	______

2. Try again.

	Guessed Sequence	Calculator Sequence
$\frac{1}{2}$	______	______
42%	______	______
0.75	______	______
$\frac{3}{8}$	______	______
65%	______	______
0.83	______	______

Lesson 18: Calculating with Percents

Overview

Students learn how to use the [%] key in computations.

Transparencies

Transparencies 12 and 13, *Percent* and *Percent Increase or Decrease*, Chapter 2

Keys Introduced

None

Student Worksheets

Sun Country
One to Another
Raises and Such
Commission er??

Teaching Steps

1. Use Transparency 12, *Percent*, to demonstrate how to use the [%] key in computation.
2. Point out how the [%] key changes the number in the display to a decimal.

Press	**Display**
25	25
[%]	0.25

3. If appropriate, demonstrate how to do some of the additional types of percentage calculations shown on Transparency 13, *Percent Increase or Decrease*.

Five-Minute Filler

A person's body is approximately 70% water. Find the weight of water in the bodies of five of your classmates.

Sun Country

Directions

1. Use the Math Explorer to find the percent each city is of the total state population of 3,131,547 people.
2. Find out what percentage the cities listed are of the total population of Arizona.

City	Population	Percent of Arizona Total
Scottsdale	115,279	______
Phoenix	816,664	______
Tempe	139,784	______
Mesa	285,689	______
Flagstaff	59,452	______
Tucson	402,191	______

One to Another

To find what percent one number is of another, express the given numbers as the numerator and denominator of a fraction. In the problem "8 is what percent of 24?" The numerator is 8 and the denominator is 24.

Directions

1. Express the numbers in the problem as a fraction.
2. Perform the indicated division or press the [FOD] key.
3. Multiply by 100 and write a % symbol.

Example: 4 is what percent of 50? $\frac{4}{50} = 0.08 \times 100 = 8\%$

a. 30 is what % of 50? ________

b. What % of 108 is 96? ________

c. 20 is what % of 50? ________

d. 50 is what % of 20? ________

e. What % of 250 is 100? ________

f. 195 is what % of 780? ________

g. 270 is what % of 900? ________

h. What % of 520 is 416? ________

i. 20 is what % of 13? ________

j. 80 is what % of 4? ________

k. What % is 78 of 109? ________

l. 245 is what % of 1000? ________

Raises and Such

Directions

Solve each of the following problems to determine the increase or decrease.

1. An air conditioner sold in a store is marked up 40% from its wholesale cost. If the wholesale cost is $370, what is the retail price?

   ____

2. Carlos makes $350 per week. If he gets a 7% raise, what is his new wage per week?

   ____

3. Norman weighed 180 pounds. After 2 months of exercising, he lost 9 pounds. What percent of his weight did he lose?

   ____

4. A coat that originally sold for $195 was marked 20% off. What is the sale price of the coat?

   ____

5. Membership in the school service club increased by 18%. If there were 54 students in the club before, what is its size now?

   ____

6. Sandi got a one-year loan for $240 for a new stereo. She had to pay 12% interest on the loan. How much was the total she had to pay to the lender?

   ____

7. By working at a certain clothing store, Robert gets a 15% discount on clothes he buys. How much can he save on $225 worth of clothes?

____

8. An investor has $29,000 invested in some bonds. The current yield on the bonds is 12% per year. How much money is she making per year?

____

9. One month $4030 worth of merchandise was returned to a shipping company because of damage in shipping. If during that period the company shipped goods worth $450,000, what percent of the merchandise was returned for damage?

____

10. Ms. Talbot purchased $44,500 worth of tax-free municipal bonds. If the interest on the bonds was 7.5% per year, how much tax-free income could she realize from the bonds each year?

____

11. The Sandpoint School District has personnel costs of $17,600,000. If $4,928,000 of the personnel costs goes to provide employee benefits, benefits are what percentage of the total personnel costs?

____

12. A sprinkler accidently discharged in a fabric store. All the fabric in the store was sold at 40% of its original cost. Alice spent $76 on the watersoaked fabric. What was the original value of the fabric?

____

Commission er??

Directions

Use the Math Explorer to solve the following problems.

1. At a boat show, a salesperson who sells a $17,500 cabin cruiser receives a commission of 15%. How much does he earn?

   ____

2. A used-car salesperson received a commission of $450 on a sale. If her commission is 17%, what was the sale price of the car?

   ____

3. The commission on the sale of a $4500 living room set was $800. What is the percentage of the commission?

   ____

4. A diamond ring that was originally priced at $750 was offered for sale at $695. What was the percentage of discount during the sale?

   ____

5. The cost per pupil of operating a school last year was $3400. This year the cost is $3750. What is the percentage of increase this year over last year?

   ____

6. Mr. Ingred's real-estate business sold 233 homes last year. This year their sales are up 9% over last year. How many homes will be sold by this company this year at this rate?

   ____

7. Mountain View school has an enrollment of 584 students this year. This is an increase of 12% over last year. How many students were there at Mountain View last year?

____

8. Shirley bought two shirts at $13.95 each. How much sales tax does she owe on the two shirts if the tax rate in her state is 7%?

____

9. Sam bought a color television set for $496. How much sales tax did he owe if the tax rate in his state is 6.5%? What was the total cost of the television set?

___________ ___________

10. If 1740 out of 3600 students are boys, what percent of the students are boys?

____

11. How much was saved by buying a tire at a reduction of 30% if the sale price was $72?

____

12. An increase of 15% in the price of an automobile amounted to $720. What was the original price?

____

13. If you owed $350 and had paid $275, what percent of your debt had you paid?

____

14. The school volleyball team won 96% of all of the games it played. If it won 48 games, how many games were played?

____

Lesson 19: Calculating Volumes and Circumferences

Overview

This lesson teaches students how to use the [x^2], [√‾], [y^x], and [π], keys. They then practice calculating volumes with the student worksheet, *The Yummy Yogurt Marketing Meeting*.

Transparencies

Transparencies 24, 25, and 27, *Powers of Numbers*, *Squares and Square Roots*, and *Pi*, Chapter 2

Keys Introduced

[y^x] [π] [x^2] [√‾]

Student Worksheets

Valid Relationships
Circle Up!
Do the Parts Equal the Whole?
The Yummy Yogurt Marketing Meeting
Square "Roots"
Pizza Bargains
Varied Dimensions

Teaching Steps

1. Use the appropriate transparencies to demonstrate how to use the [x^2], [y^x], [√‾], and [π] keys. Point out that the sequence with the [y^x] key is: (base) [y^x] (exponent) [=]

2. Review these formulas:

   - Circumference of a circle: $C = 2\pi r = \pi d$
   - Area of a circle: $A = \pi r^2$
   - Volume of a cylinder: $V = \pi r^2 h$
   - Volume of a sphere: $V = \frac{4}{3}\pi r^3$
   - Volume of a cone: $V = \frac{1}{3}\pi r^2 h$

Five-Minute Filler

If you have a cube that is 2 centimeters on each side, the volume is 8 cubic centimeters. Its surface area is 24 square centimeters. If you cut the cube down the middle each way so that you have eight cubes, 1 centimeter on each side, what is the new total volume and surface area?

Valid Relationships

Directions

1. Collect the objects listed in the chart.
2. Find the relationship between the circumference and the diameter of each object.
3. Complete the chart. *C* stands for circumference and *D* stands for diameter.

Object	C	D	C ÷ D	C + D	C - D	C × D
Pop can	_____	_____	_____	_____	_____	_____
Coffee can	_____	_____	_____	_____	_____	_____
Plate	_____	_____	_____	_____	_____	_____
Glass	_____	_____	_____	_____	_____	_____
Ring	_____	_____	_____	_____	_____	_____
Bike wheel	_____	_____	_____	_____	_____	_____

4. Which column shows a constant relationship between the circumference and the diameter?
5. Find the average for that column and compare that average with the known value of π (3.14).

Circle Up!

Directions

1. For each of the circles, fill in the blanks.

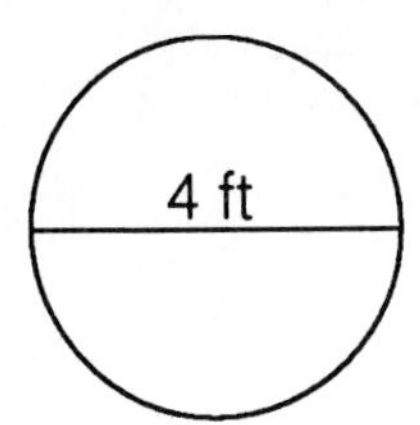

a. $C =$ ______

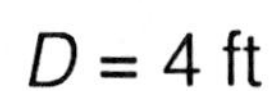

$D = 4$ ft

$r =$ ______

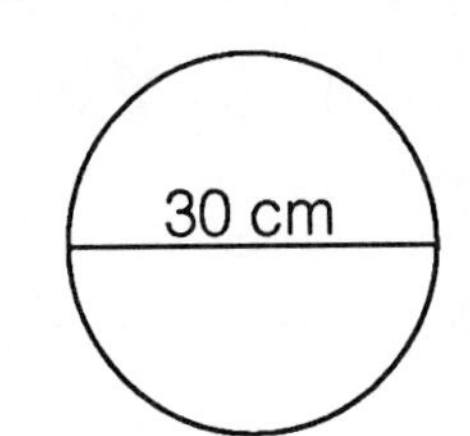

b. $C =$ ______

$D = 30$ cm

$r =$ ______

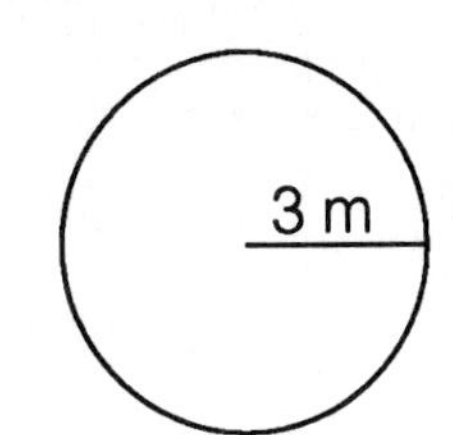

c. $C =$ ______

$D =$ ______

$r = 3$ m

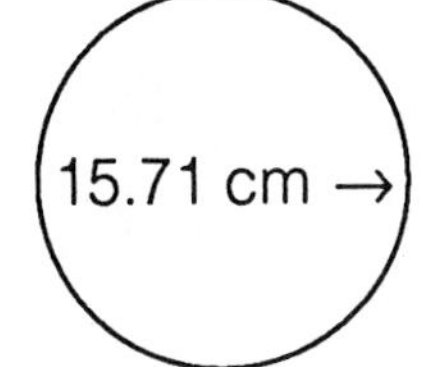

d. $C = 15.71$ cm

$D =$ ______

$r =$ ______

37.7 m →

e. $C = 37.7$ m

$D =$ ______

$r =$ ______

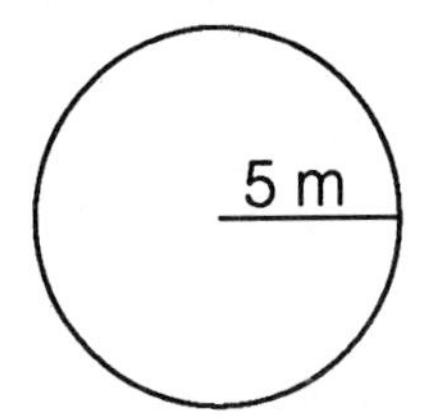

f. $C =$ ______

$D =$ ______

$r = 5$ m

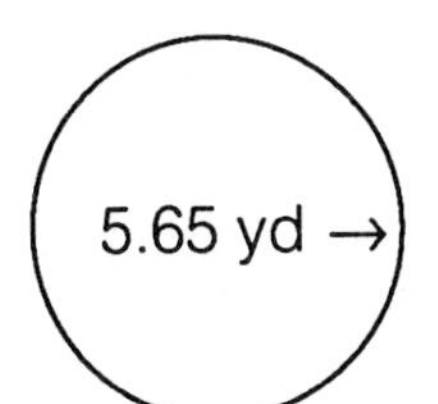

g. C = 5.65 yds

$D =$ ______

$r =$ ______

72.26 in. →

h. $C = 72.26$ in.

$D =$ ______

$r =$ ______

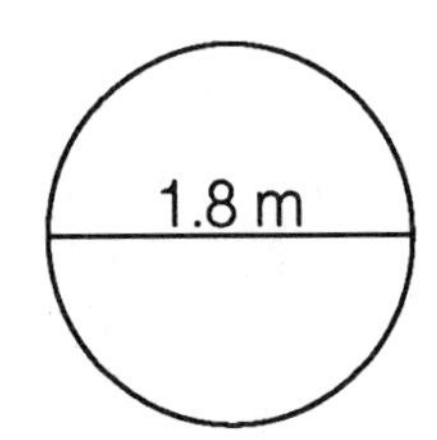

i. $C =$ ______

$D = 1.8$ m

$r =$ ______

2. Find the distance around each figure.

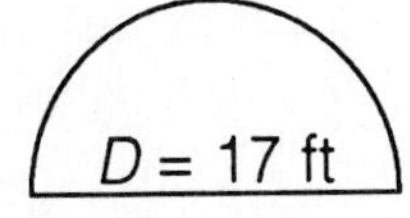

j. ____________

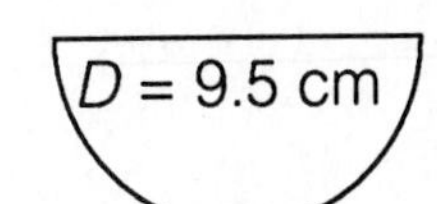

k. ____________

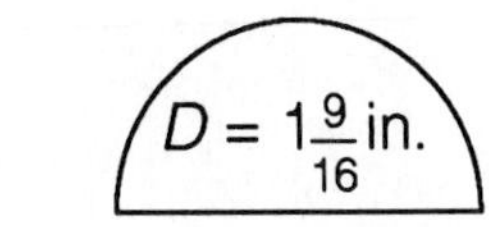

l. ____________

Do the Parts Equal the Whole?

Directions

1. To find the area of a circle, use a round piece of cardboard.
2. Cut the lid so the sections fit together as shown.

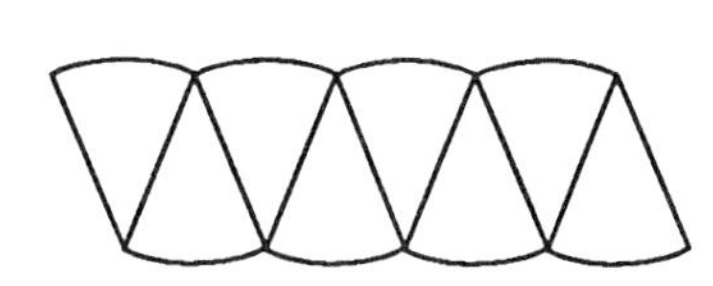

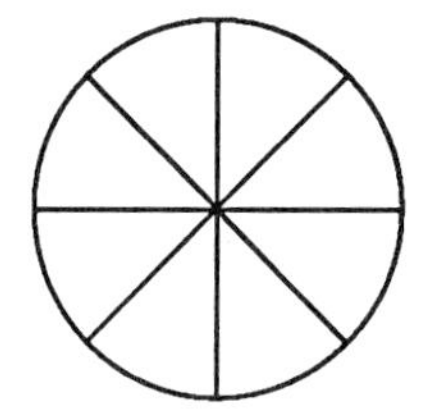

The area of the circle approaches that of a parallelogram with a base of $\frac{1}{2}C$, or πr, and an altitude of r. Thus the area of the circle is πr^2.

3. Use the Math Explorer to verify the approximate value by using the formula $A = \pi r^2$.

Historical Note:

4. Ancient Egyptians found an approximation of the area of a circle by finding the area of a square whose side is $\frac{8}{9}$ of the diameter of the circle.
5. Try the Egyptian method on the two circles below.

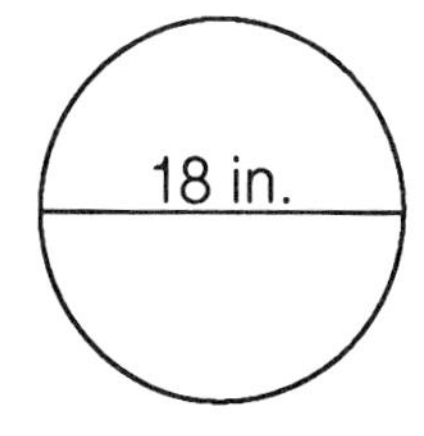

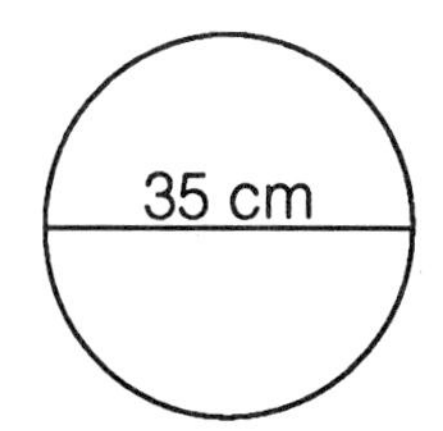

6. Check these approximations by using the formula for the area of a circle.

The Yummy Yogurt Marketing Meeting

The Situation:

The Yummy Yogurt Company is trying to decide on a container for its brand new shrimp-flavored yogurt, "Surprisingly Shrimp." They want a package that is unique and large enough. The marketing department has suggested the containers below. Which should the company choose?

1. Calculate the volume of each container to hundredths.

Formulas Cylinder: $V = \pi r^2 h$

Sphere: $V = \frac{4}{3} \pi r^3$

Cone: $V = \frac{4}{3} \pi r^2 h$

(Container *D*: cylinder + cone)

2. Rank the containers by volume with 1 for the largest and 4 for the smallest.

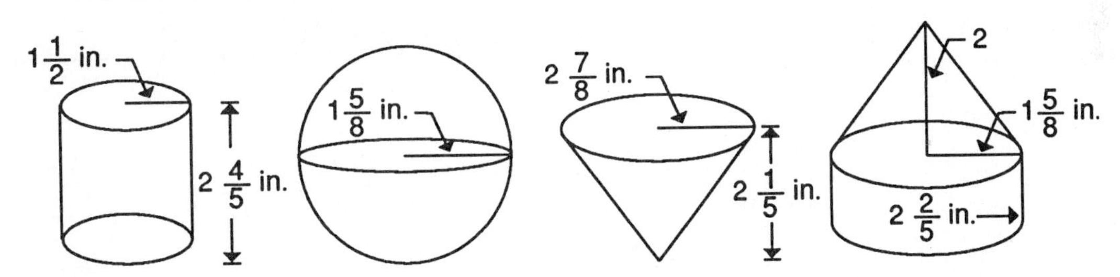

Container A **Container B** **Container C** **Container D**

Container	Volume	Rank by Volume
A	________	________
B	________	________
C	________	________
D	________	________

3. The company also wants to find out if anyone will buy "Surprisingly Shrimp." Put a check in the box that best expresses your opinion about this new product.

   - ☐ I would pay lots of money for shrimp-flavored yogurt.
   - ☐ I would buy this yogurt instead of spinach.
   - ☐ I don't like shrimp.

Square "Roots"

Directions

1. Complete the following chart.
2. If only the area is given, to find the radius you will have to:
   a. Divide the area by π.
   b. Find the square root of the quotient.

Area	Radius	Diameter
379.98 cm^2		
256.78 cm^2		
	8.4 m	
385 cm^2		
3215.36 cm^2		
		625.47 mm
		47.5 cm
	$\frac{1}{3}$m	

Pizza Bargains

Directions

Determine which buy is the best one.

1. Do you get more pizza if you buy two 20-centimeter pizzas or one 40-centimeter pizza?

   ____

2. Do you get more pizza if you buy two 30-centimeter pizzas or one 40-centimeter pizza?

   ____

3. Do you get more pizza if you buy two 20-centimeter pizzas or one 30-centimeter pizza?

   ____

4. If a 20-centimeter pizza is priced at $3.95, determine a fair price for a 30-centimeter pizza.

   ____

5. If a 20-centimeter pizza is priced at $3.95, determine a fair price for a 40-centimeter pizza.

   ____

Varied Dimensions

Directions

1. Use the cylinder shown to answer the following questions. Use the Math Explorer to find the volume of the cylinder if:

   a. The height is doubled. __________

   b. The radius is doubled. __________

   c. The radius and the height are doubled. __________

   d. The height is halved. __________

   e. The radius is halved. __________

   f. The height and radius are halved. __________

   g. The radius and the height are multiplied by 5.

   ____

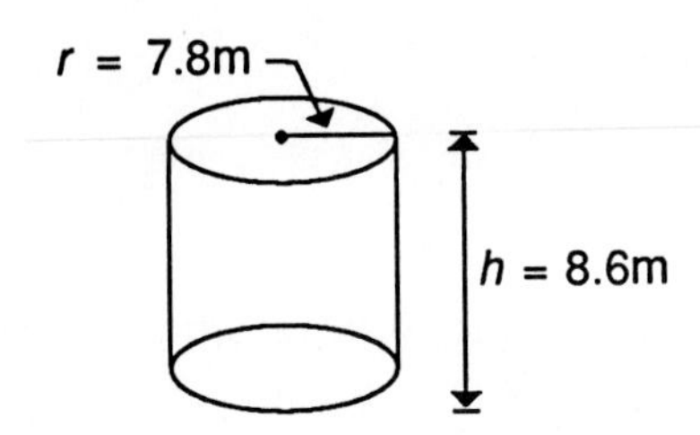

Lesson 20: Raising Numbers to a Power

Overview

Students learn to find the square root of a number and raise a number to a power.

Transparencies

Transparencies 24 and 25, *Powers of Numbers* and *Squares and Square Roots,* Chapter 2

Keys Introduced

None

Student Worksheets

Approximations
Which Library is This?
Who was Pythagoras?
Power Match

Teaching Steps

1. Use Transparencies 24 and 25, *Powers of Numbers* and *Squares and Square Roots,* to demonstrate use of keys $\boxed{y^x}$ and $\boxed{x^2}$.
2. Have the students work all the problems in the worksheet *Approximations* before they use the square root key.
3. Have the students complete the appropriate worksheets in this lesson.

Five-Minute Fillers

1. The square root of 64 is 8, and the cube root of 64 is 4. Find another number that has a whole-number square root and cube root.
2. Third powers of numbers can end in any digit from 0 through 9. Is this true for fourth powers of numbers? Is it true for fifth powers of numbers?

Approximations

The square root of a number *n* is defined as the number that equals *n* when multiplied by itself.

Directions

Find the square root of 985 by following the steps listed.

1. Determine two squares between which the number falls.

   $30 = \sqrt{900}$ $40 = \sqrt{1600}$

2. Estimate an approximate square root: 31

3. Divide and round: $985 \div 31 = 31.77$

4. Find the average of the divisor and the quotient:

   $$\frac{31 + 31.77}{2} = 31.385$$

5. Divide by the average:

   $985 \div 31.38 = 31.384$

6. Repeat the averaging process if more accuracy is desired.

7. Find the square root of each number to the nearest hundredth.

   a. 789 ________ d. 1045 ________

   b. 125 ________ e. 4567 ________

   c. 56 ________ f. 34,788 ________

9. Check your answers to the problems by entering the number in the Math Explorer and pressing [√].

Which Library is This?

Directions

1. Answer the question "What book contains 66 books?" by using the Math Explorer, where you can, to perform the indicated operations.
2. Once each answer is obtained, turn the Math Explorer upside down to read the clues.
   a. Start with the square root of 484.
   b. Multiply that answer by the square root of 5184.
   c. Multiply the answer in (b) by the square root of 36,481.
   d. Now divide by the square root of 64.

   ____

Who was Pythagoras?

Directions

1. Below is an illustration of the Pythagorean theorem. The theorem states that if squares are drawn on the three sides of a right triangle, the sum of the squares of the number of squares on each of the two shorter sides will be equal to the square of the number of squares on the long side:

   $a^2 + b^2 = c^2$.

2. Use the Math Explorer to determine whether the following triangles are right triangles.

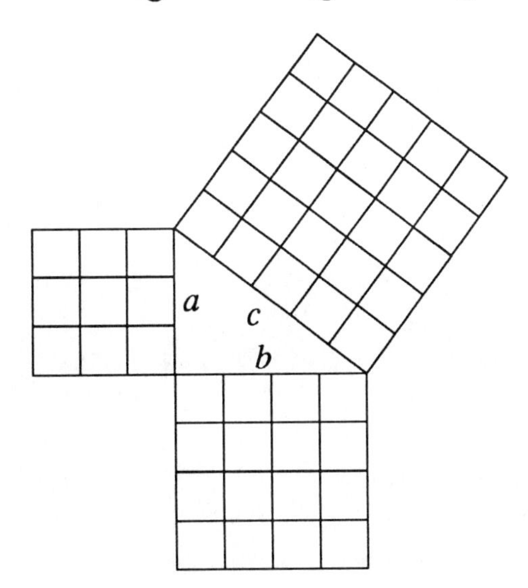

	a	b	c	
a.	81	144	225	______
b.	4	5	5	______
c.	36	64	100	______
d.	40	30	50	______
e.	12	16	25	______
f.	9	12	13	______
g.	15	20	35	______
h.	15	8	17	______
i.	7	24	25	______
j.	10	24	26	______
k.	8	13	16	______

Power Match

Directions

1. Try to find the power to match each problem with its correct answer.
2. Work the problems to find the power to match an answer on the Math Explorer.
3. Correct any errors you may have made.

a. 23 — = ________ 441

b. (0.3) — = ________ 0.8957046

c. 2 — = ________ 279,841

d. 21 — = ________ 0.00243

e. 3 — = ________ 64

f. (0.032) — = ________ 6561

Try Again!

a. 19 — = ________ 0.3689348

b. (0.07) — = ________ 0.000343

c. 6 — = ________ 4,826,809

d. (0.6074) — = ________ 59,049

e. 9 — = ________ 2,476,099

f. 13 — = ________ 46,656

Lesson 21: Reciprocals

Overview

Students learn to use the calculator to find the reciprocal of any given whole number, fraction, or mixed number.

Transparency

Transparency 15, *Reciprocals*, Chapter 2

Key Introduced

[1/x]

Student Worksheets

Upside Down
Over or Under?
One Time!
Increased Dividends

Teaching Steps

1. Use Transparency 15, *Reciprocals*, to demonstrate how a reciprocal is obtained.
2. Discuss the following concept: For every number not equal to 0, there is a nonzero number, called its reciprocal, or inverse for multiplication, such that the product of the two numbers is 1.
3. Students need to know that division by zero is undefined, but that is not the important idea with this set of exercises. Use the activity ***Increased Dividends*** to help students visualize this concept if you feel it is an appropriate topic.

Five-Minute Filler

What is the smallest number you can enter into the display of the Math Explorer? What is the largest number you can enter into the display of the Math Explorer, and when pressing the [1/x] key, not get an error message?

Upside Down

Directions

1. Enter each number into the display of the Math Explorer.
2. Press the [1/x] key and then the [F⇔D] key. Simplify if possible. Record your answer by the number.

a. 100 _____	d. 2 _____	g. $\frac{1}{5}$ _____
b. 25 _____	e. 1 _____	h. $\frac{1}{25}$ _____
c. 5 _____	f. $\frac{1}{2}$ _____	i. 0.01 _____

3. Write a rule that describes the relationship between the numbers listed and their reciprocals.

____

4. How would you write the reciprocal of 0?

____

Over or Under?

Directions

Write the reciprocal for each number. Use the Math Explorer to check your answer.

a. 12 ______

b. $\frac{1}{9}$ ______

c. $2\frac{1}{2}$ ______

d. 6 ______

e. $\frac{4}{7}$ ______

f. $3\frac{5}{8}$ ______

g. 52 ______

h. $\frac{7}{100}$ ______

i. $4\frac{2}{3}$ ______

j. 10,000 ______

k. $\frac{1}{64}$ ______

l. $10\frac{1}{3}$ ______

m. 28 ______

n. $\frac{9}{11}$ ______

o. $23\frac{2}{3}$ ______

p. 195 ______

q. $\frac{47}{64}$ ______

r. $90\frac{19}{36}$ ______

One Time!

Directions

Given that [ ] × { } = 1, complete the table.

	[ ]	{ }
a.	1	______
b.	2	______
c.	$\frac{1}{2}$	______
d.	$\frac{1}{3}$	______
e.	$\frac{1}{4}$	______
f.	$\frac{2}{3}$	______
g.	______	$1\frac{1}{4}$
h.	______	$1\frac{1}{8}$
i.	$\frac{1}{5}$	______
j.	$\frac{3}{4}$	______
k.	______	$\frac{2}{7}$
l.	______	$1\frac{3}{5}$
m.	$\frac{3}{8}$	______

Increased Dividends

Directions

1. Enter a small number into the display of your calculator.

2. Press the [1/x] key. Record your answer.

   ____

3. Divide the number 1 by that same small number. Compare the answer with your answer in step 2.

   ____

4. Enter an even smaller number into the display of your calculator and find its reciprocal.

   ____

5. Keep doing this until you have entered the smallest number you can possibly enter into your calculator.

6. Describe in your own words what happened to the reciprocals as the numbers you entered into the display of your calculator got smaller.

   ____

7. Since finding the reciprocal is the same as dividing 1 by the number, dividing by numbers that approach 0 would cause the quotient to do what?

   ____

Lesson 22: Using Powers of 10

Overview

Students learn how to use the powers of 10 key.

Transparencies

Transparency 23, *Powers of 10*, Chapter 2

Keys Introduced

[10ⁿ]

Student Worksheets

Notable Notations

Teaching Steps

1. Use Transparency 23, *Powers of 10*, to demonstrate how to use the [10^n] key.
2. Point out that the [10^n] key can be used to change a number from scientific notation to decimal notation. For 5.6×10^4 and 0.000078×10^5:

Press		Display	Press		Display
5.6		5.6	0.000078		0.000078
[×]	×	5.6	[×]	×	0.000078
[10^n]	×	5.6	[10^n]	×	0.000078
4	×	10000	5	×	100000
[=]		56000	[=]		7.8

3. Have students work through the chapter on scientific notation as part of this lesson.

Five-Minute Filler

If there are 2.6×10^8 people in the United States and you could count one person per second, how long would it take you to count all the people in the United States?

Notable Notations

Directions

1. Write each of the following problems in decimal notation.
2. Check your answers by working the problems on the Math Explorer.

a. 5×10^{4} ____________

b. 4.5×10^{-3} ____________

c. 3.6×10^{2} ____________

d. 7.3×10^{-2} ____________

e. 9×10^{5} ____________

f. 1.7×10^{-5} ____________

g. 2.023×10^{2} ____________

h. 9.01×10^{-4} ____________

i. 5.04×10^{4} ____________

j. 6.02×10^{-5} ____________

k. 1.01×10^{2} ____________

l. 4.033×10^{-2} ____________

3. Use the memory keys to solve the following problems.

$(2.5 \times 10^{2})(3 \times 10^{3}) =$ ____________

$\frac{42 \times 10^{4}}{7 \times 10^{-1}} =$ ____________

$(9.46 \times 10^{4})(1.50 \times 10^{-5}) =$ ____________

$\frac{1.6 \times 10^{-2}}{4 \times 10^{-4}} =$ ____________

$\frac{1.08 \times 10^{-3}}{(2 \times 10^{2})(3 \times 10^{-3})} =$ ____________

$\frac{16 \times 10^{-2}}{16 \times 10^{2}} =$ ____________

$(1.63 \times 10^{4})(9.81 \times 10^{-3}) =$ ____________

$\frac{6.55 \times 10^{-3}}{9.3 \times 10^{-3}} =$ ____________

Lesson 23: Signed Numbers

Overview

Students learn how to use the change-sign key.

Transparency

Transparency 14, *Changing the Sign of a Number*, Chapter 2

Keys Introduced

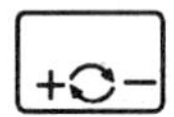

Student Worksheets

Learn the Signs
More Signs to Follow
Signs of the Times
Signed On

Teaching Steps

1. Use Transparency 14, *Changing the Sign of a Number*, to demonstrate how to use the [+⇄-] key.
2. Point out how the [+⇄-] key changes the sign of the number in the display without changing the operation the calculator is to perform. For 91 ÷ (-7):

Press	Display
91	91
[÷]	÷ 91
7	7
[+⇄-]	-7
[=]	-13

3. Have the students complete the appropriate worksheets.

Five-Minute Filler

What is the greatest possible value that can be obtained by multiplying a number from 0 to 9 by a number between -5 and 0?

Learn the Signs

Directions

1. Enter the first number of the problem in the Math Explorer. If this number is negative, press the [+○-] key.
2. Press the operation symbol key.
3. Enter the second number in the Math Explorer. If the second number is negative, press the [+○-] key.
4. Press the [=] key.

   Note: The parentheses are used to separate parts of the problem and should not be entered into the Math Explorer.

   a. (+9) + (-5) _______ e. (-23) + (+15) _______

   b. (-9) + (-5) _______ f. (-17) + (-234) _______

   c. (-9) + (+5) _______ g. (+412) + (-73) _______

   d. (-23) + (-15) _______ h. (-89) + (-111) _______

5. Look closely at the problems you have just worked on the Math Explorer. Circle the word in parentheses that makes the statement true.

   a. If the signs of the numbers are different, (add/subtract) and give the answer the sign of the (larger/smaller) number.

   b. If the signs are the same, (add/subtract) and give the answer which sign?

   ____

6. Work the following problems mentally and then check your answers with the Math Explorer.

   a. (+6) + (-4) + (-7) _____ d. (-3) + (-8) + (-2) _____

   b. (+12) + (-5) + (+ 3)_____ e. (+6) + (-2) + (+1) _____

   c. (-5) + (-4) + (+9) _____ f. (+6) + (-5) + (+3) + (-9)_____

More Signs to Follow

Directions

1. Work the following problems on the Math Explorer.

2. Pay particular attention to the sign the Math Explorer assigns to each answer.

   a. (+8) – (+6) ______
   b. (-8) – (+6) ______
   c. (+12) – (+8) ______
   d. (-7) – (+3) ______
   e. (-8) – (-6) ______
   f. (+9) – (-6) ______
   g. (-12) – (-8) ______
   h. (+15) – (-7) ______
   i. (-18) – (-20) ______
   j. (-54) – (+6) ______
   k. (-23) – (-30) ______
   l. (+17) – (-3) ______
   m. (+42) – (+12) ______
   n. (+18) – (-13) ______

3. Complete the following statement about how to solve subtraction problems involving signed numbers:

   To subtract signed numbers, change the sign of the ______________ and follow the rules of ______________ for signed numbers.

4. Work the following problems mentally. Check your answers by working the problems on the Math Explorer.

   a. (+7) - (+4) _____
   b. (-10) - (-4) _____
   c. (-6) - (+6) _____
   d. (-30) - (-29) _____
   e. (+18) - (-9) _____
   f. (+56) - (+16) _____
   g. (-6) - (+5) - (-2) - (-1) _____
   h. (+6) - (-2) - (+7) - (-4) _____
   i. (+2) - (-4) + (-3) - (+1) _____
   j. (4) + (-5) - (7) + (2) _____

Signs of the Times

Directions

1. Work the following problems on the Math Explorer.
2. Pay particular attention to the answer the Math Explorer gives and the signs of the numbers in the problem.

   Note: If there are no signs between sets of parentheses, multiplication is the implied operation.

   a. (+6)(-2) ______ e. (-45)(+65) ______
   b. (-9)(-3) ______ f. (+23)(+78) ______
   c. $(+\frac{1}{3})(-12)$ ______ g. (-90)(-87) ______
   d. (-12)(-17) ______ h. (-123)(+45) ______

3. Complete the following statement about how to solve multiplication problems involving signed numbers.

   To multiply signed numbers, if the numbers are the same sign, the answer is ______________. If the numbers have differ-ent signs, the answer is ______________.

4. Work the following problems mentally. Check your answers by working the problems on the Math Explorer.

   a. (-3)(+4) ______ d. (+7)(-2)(-5) ______
   b. (-2)(-7) ______ e. (-16)(+2)(-3) ______
   c. (-5)(-2)(+3) ______ f. (+12)(-2)(+5) ______

5. Try the following problems on the Math Explorer. How do the rules for division of signed numbers compare to the rules for multiplication? ________________________________

   a. 24 ÷ (-6) ______ d. (-6)(+5) ÷ (-3)(+2) ______
   b. (-20) ÷ (-4) ______ e. (-11)(19) ÷ (-14) ______
   c. $(-6) \div (\frac{1}{3})$ ______ f. (19)(-7) ÷ (-13) ______

Signed On

Directions

Solve the following problems using the Math Explorer.

1. During one week of a recent winter the following temperatures, in degrees Celsius, were registered in Antarctica:

   -81 -72 -79 -84 -78 -80 -74

   What was the average temperature that week?

   ______

2. The booster on a model rocket detached at 400 feet. How long did it take to reach the ground?

   The formula for finding the time is $t = \sqrt{\frac{2s}{a}}$ where s is the distance the booster falls (-400 feet) and a is the acceleration due to gravity (-32 feet per second per second).

   ______

3. The Pythagorean theorem for finding the third side of a right triangle when two sides are known is

   $a^2 + b^2 = c^2$

   If $b = 30$ feet and $c = 34$ feet, a can be found by the formula

   $\sqrt{a} = \sqrt{c^2 - b^2}$

   How long is a? ____________

Lesson 24: Geometry Applications

Overview

Students will use the operations available on the Math Explorer to solve geometry application problems.

Transparencies

None

Keys Introduced

None

Student Worksheets

Estimate the Area
Roaming Areas
Long John Silver?
Prison or Prism?

Teaching Steps

1. If there is a need to review some of the key functions, use the appropriate transparencies from previous lessons.
2. Be sure that students have an understanding of the formulas needed to solve the problems.
3. Have students do worksheets appropriate for their ability.

Five-Minute Filler

Complete the chart that shows the number of line segments that can be associated with a given number of points:

Points	2	3	4	5	6	7
Segments	1	3	6	__	__	__

Estimate the Area

Directions

Estimate the area first. Then, using the Math Explorer calculator, find the area of each of the drawings.

1.

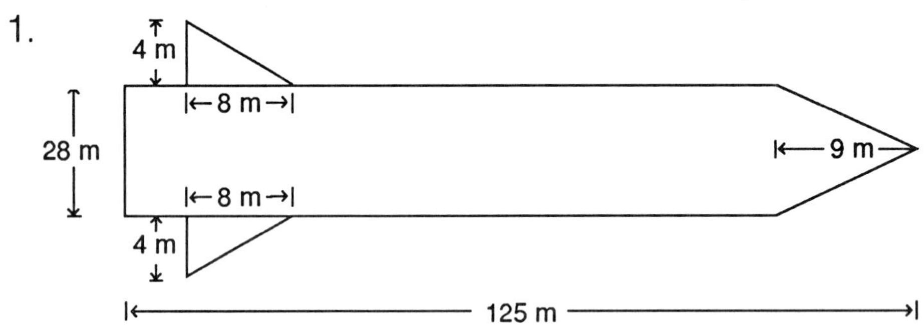

Your estimate __________ Area __________

2.

10 cm
15 cm
10 cm
72 cm
15 cm
8 cm
10 cm
10 cm
10 cm

Your estimate __________ Area __________

3.

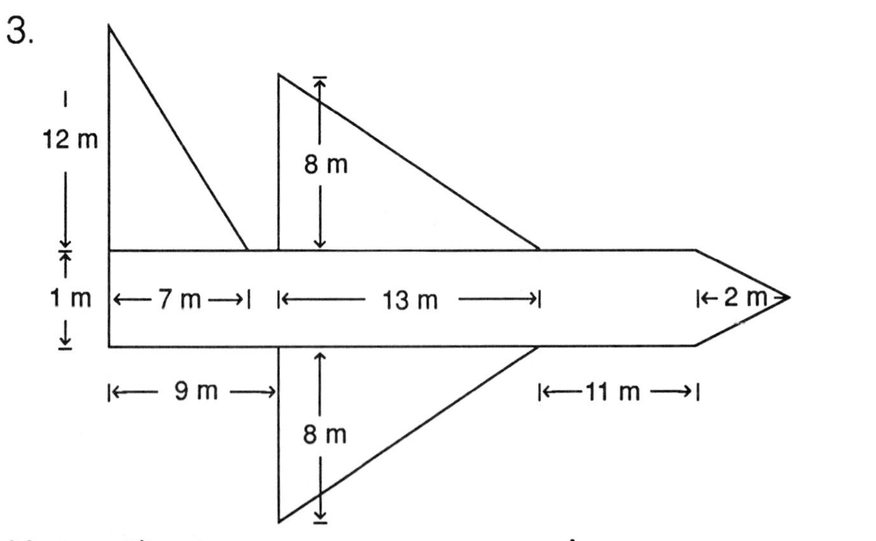

Your estimate __________ Area __________

Roaming Areas

Directions

1. Use the Math Explorer and the formula for finding the area of a trapezoid,

$$A = \frac{(B + b)\,h}{2}$$

to find the area of each of the trapezoids shown.

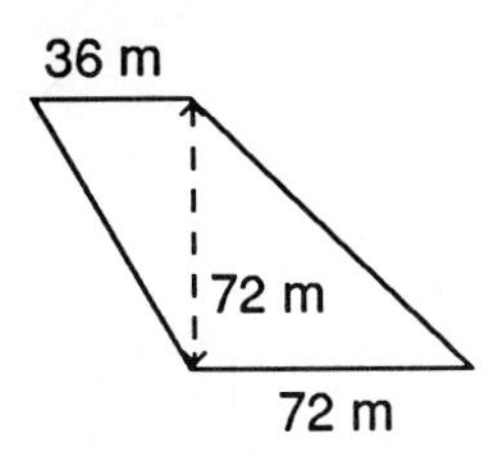

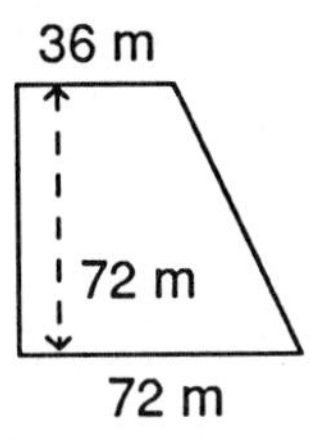

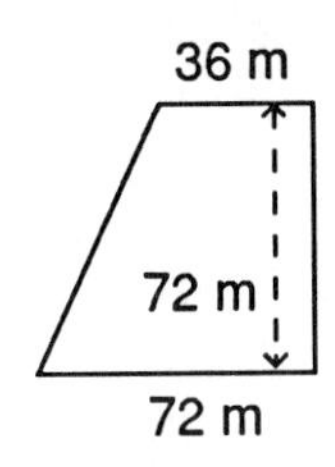

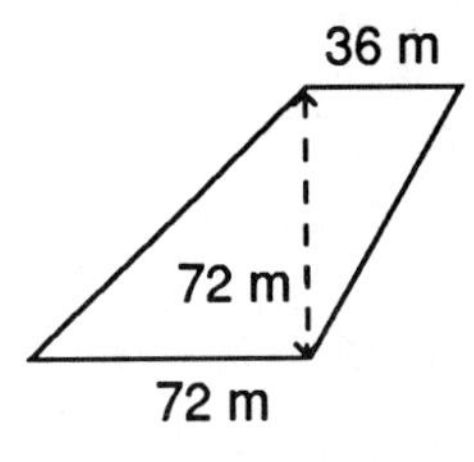

a. ________ b. ________ c. ________ d. ________

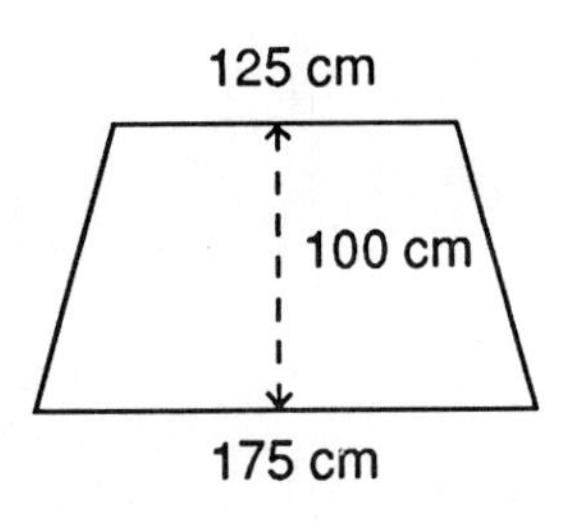

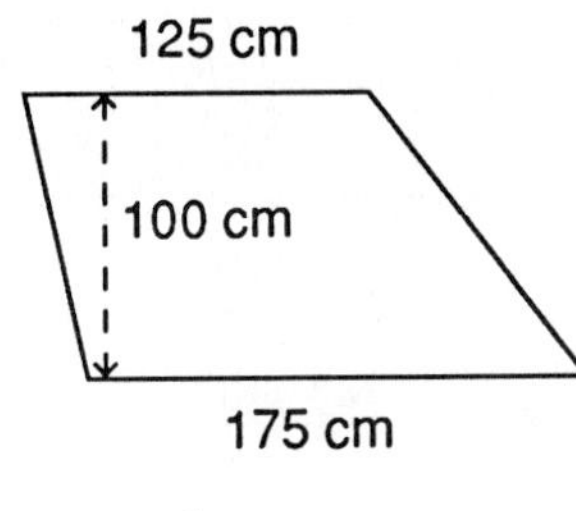

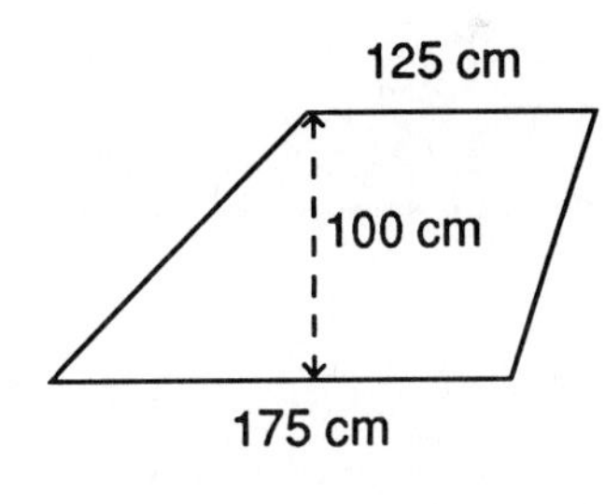

e. ________ f. ________ g. ________

2. Explain your answers.

____

____

Long John Silver?

Directions

1. Use the Math Explorer and formula for finding the area of a triangle,

$$A = \frac{b \times h}{2}$$

to find the value of the missing variable.

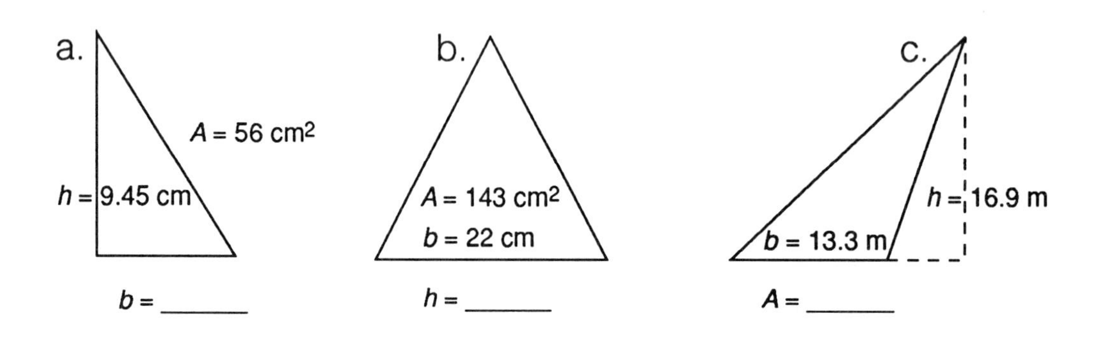

d.

$A = 508\ cm^2$

$b = 121$ cm

$h =$ ______

Prison or Prism?

Directions

1. Use the formula for finding the volume of a prism:

   V = area of the base times the height

2. Use the Math Explorer to find the volume of the prism.

   V = ______

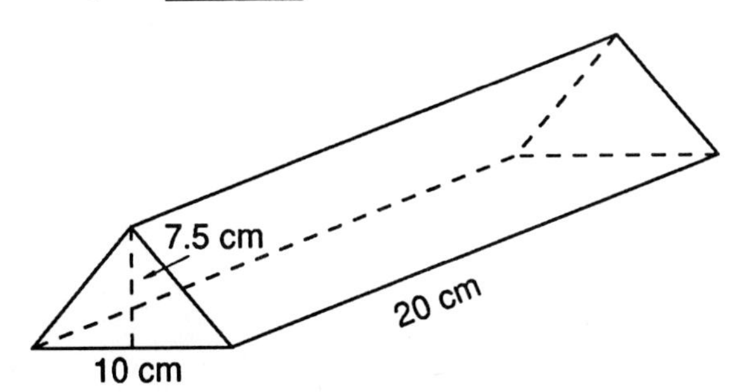

3. Tell how the volume changes for the prism if:

   a. The length is doubled.

   ____

   b. The area of the base is doubled.

   ____

   c. The length, width, and height are all doubled.

   ____

   d. The height is halved.

   ____

   e. The length, width, and height are all halved.

   ____

   f. The length, width, and height are all multiplied by 3.

   ____

Lesson 25: Statistics

Overview

Students will use the operations available on the Math Explorer to solve statistics application problems.

Transparencies

None

Keys Introduced

None

Student Worksheets

Average Magic
North to the Future?
Mean, Median and Mode
How Frequently Does It Occur?

Teaching Steps

1. If there is a need to review some of the key functions, use the appropriate transparencies from previous lessons.
2. Have students do worksheets appropriate for their ability.
3. Help students collect statistics from their everyday experiences and perform the appropriate statistical analyses using them.

Five-Minute Filler

Read this to the class from Mark Twain's "Life on the Mississippi," published in 1901.

"In the space of one hundred and seventy-six years the Lower Misissippi has shortened itself two hundred and forty-two miles. That is an average of a trifle over one mile and a third per year. Therefore, any calm person, who is not blind or idiotic, can see that in the Old Oolithic Silurian Period, just a million years ago last November, the Lower Mississippi River was upward of one million three hundred thousand miles long, and stuck out over the Gulf of Mexico like a fishing rod.

And by the same token any person can see that seven hundred and forty-two years from now the Lower Mississippi will be only a mile and three-quarters long, and Cairo and New Orleans will have joined their streets together, and be plodding comfortably along under a single mayor and a mutual board of aldermen. There is something fascinating about science. One gets such wholesome returns of conjecture out of such a trifling investment of fact."

Average Magic

Directions

Use the Math Explorer to determine whether the mean (average) in each section yields a value that allows the square to be a magic square, i.e., all rows, columns, and diagonals have the same sum.

159.35 217.98 149.73 230.38	16.93 36.32 18.71 19.15 27.24	123.63 217.14 219.67 91.68 133.37 66.63
78.60 52.93 45.12 71.25 68.39 88.88 91.90	69.75 78.96 45.69 147.32 250.03	123.45 189.79 183.83
78.69 121.23 81.56 99.99 87.78 98.83	258.63 187.86 203.29 202.34	45.21 51.58 44.36 39.78 63.11 45.79 37.80 51.09

North to the Future?

Directions

Calculate the mean for each set of Alaska statistics.

1. What is the average number of visitors to Alaska over the 10-year period shown in the chart?

Year	Number of Visitors
1977	505,200
1978	522,500
1979	546,000
1980	570,600
1981	596,600
1982	623,100
1983	646,000
1984	672,000
1985	700,000
1986	787,000

Average: __________

2. Find the average number of snowmobile registrations in Alaska during the years shown in the chart.

Year	Snowmobile Registrations
1977	3797
1978	3539
1979	4618
1980	5213
1981	5426
1982	5879
1983	6114
1984	6217

Average: __________

Mean, Median and Mode

The median of a distribution is its middle value.
The mode of a distribution is the most frequently occurring value.

Directions

1. Arrange the data in ascending or descending order.
2. Find the median, mode, and mean of each set of data.

a. 42, 51, 41, 36, 50, 38, 43

__________ __________ __________

b. At ACME Steel Company, the salaries for the office employees are as follows:

$13,500, $17,800, $19,450, $17,800, and $54,000

__________ __________ __________

c. Find the median and the mean of the data in the *X* column at the right.

__________ __________

d. Find the median and the mean of the data in the *Y* column at the right.

__________ __________

e. Suppose the data in the X column represents the price of a share on the stock exchange and Y represents the number of shares sold at that price on a given day. What was the mean price of a share that day? __________

X	*Y*
1.750	525
6.250	185
4.375	286
9.875	235
4.250	235
6.500	255
3.750	685
5.625	785
8.875	755

How Frequently Does It Occur?

Directions

1. Consider a situation where different values of a measurement occur more than once.
2. Obtain the arithmetic mean (average) by multiplying the value of each measurement by the number of times it occurs.
3. Add these products.
4. The mean is determined by dividing the product total by the total number of occurrences.
5. Use the Math Explorer to complete the chart and find the arithmetic mean.

Measured Value	Occurrence	Product
18	1	18
17	2	34
16	22	
15	13	
14	32	
13	15	
12	23	
11	32	
10	14	

Totals ___________ ___________

Mean ___________

Lesson 26: Special Challenges

Overview

Students will use the operations available on the Math Explorer to solve the challenge problems.

Transparencies

None

Keys Introduced

None

Student Worksheets

Millionaires!
Building Fences
Body Works!
Math Explorer Overflow
Only Two Keys Fit
Did You Get That?
Check Your Answers!
Calendar Magic
Temperature Conversion

Teaching Steps

1. If there is a need to review some of the key functions, use the appropriate transparencies from previous lessons.
2. Have students do worksheets appropriate for their ability.
3. Allow students to challenge themselves, to work as groups, and to compare answers and processes, but do not let them get frustrated if the problems are too difficult for them.

Five-Minute Fillers

1. The number of possible combinations of objects or events is given by N!. If N = 4, then $N! = 4 \times 3 \times 2 \times 1$. Check this out by determining the number of ways you can arrange the following groups of letters: *AB, ABC, ABCD.* Then compare your results with 2!, 3!, and 4!.

Improper Changes

Mixed Number	Improper Fraction	Mixed Number	Improper Fraction
	$\frac{23}{4}$	8	
$1\frac{25}{64}$			$\frac{237}{14}$
	$\frac{29}{6}$		$\frac{396}{23}$
$7\frac{2}{17}$		$12\frac{1}{3}$	
$5\frac{5}{9}$			$\frac{291}{13}$
	$\frac{295}{17}$	$12\frac{2}{7}$	
	$\frac{101}{10}$		$\frac{48}{11}$
$3\frac{1}{6}$			$\frac{1053}{29}$
$10\frac{1}{10}$		$7\frac{3}{7}$	

Lesson 14: Mixed Number Operations

Five-Minute Filler

$25\frac{1}{2} \times 32\frac{1}{4}$ inches $\quad A = 822\frac{3}{8}$ inches^2

Another Mixed Bag

a. $1\frac{3}{4}$

b. 3

c. $\frac{3}{8}$

d. $2\frac{26}{35}$

e. $62\frac{1}{2}$

f. $7\frac{3}{4}$

g. $\frac{25}{48}$

h. $1\frac{1}{2}$

i. $4\frac{7}{12}$

j. $12\frac{1}{3}$

k. $3\frac{11}{12}$

l. $2\frac{32}{35}$

m. $10\frac{5}{7}$

n. $10\frac{1}{9}$

o. $\frac{31}{40}$

15 plates

The Trouble Maker

a.b.c.	5,1,4 (hIS)
d.e.f.g.h.	5,5,5,0,8 (BOSS'S)
i.j.k.l.m.	5,3,0,4,5 (ShOES)

Is It Really Magic?

$1\frac{1}{3}$	$\frac{1}{2}$	
$\frac{1}{6}$		
	$1\frac{1}{6}$	$\frac{1}{3}$

Yes!

		5
	$3\frac{7}{8}$	
	$5\frac{3}{8}$	$1\frac{1}{2}$

No!

(This answer area is blank.) No!

	$\frac{5}{6}$	$6\frac{2}{3}$
$5\frac{5}{6}$		$2\frac{1}{2}$
	$7\frac{1}{2}$	

Yes!

The Ancients of Noitcarf

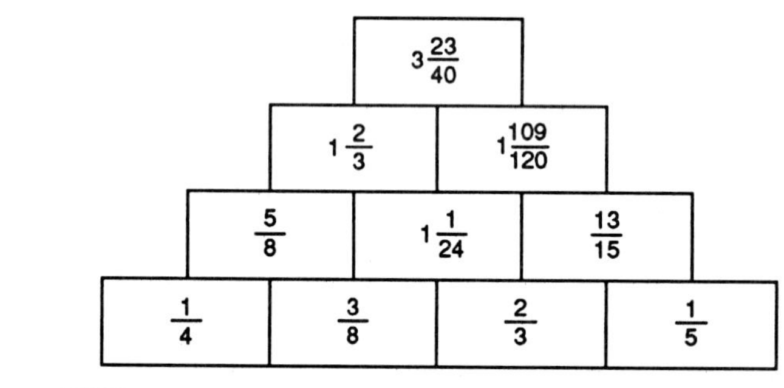

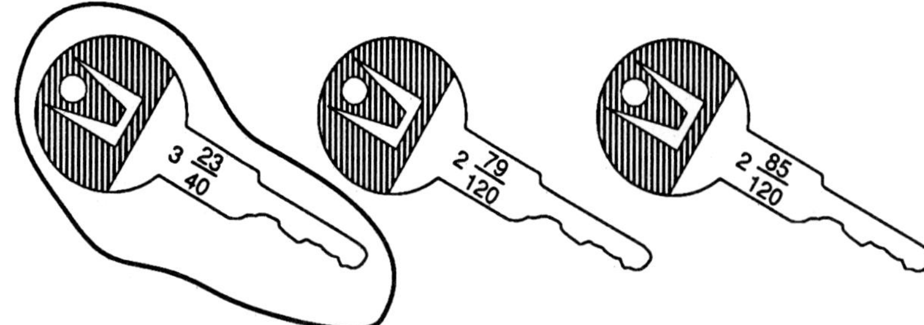

Lesson 15: Operations on Decimals

A Record Event

3. a. 1, 1 actual: 20.28
   b. 2, 2 actual: 204.1344
   c. 2, 1 actual: 24.099
   d. 3, 1 actual: 0.3397
   e. 2, 0 actual: 48.36
   f. 3, 2 actual: 0.10947
   g. 1, 3 actual: 253.6794
   h. 1, 3 actual: 997.9002
   i. 3, 2 actual: 0.02916

A Record Event (cont.)

4. The sum of the decimal places in the factors gives the number of decimal places in the answer.

Determination

Magic Square

e. 6	b. 1	h. 8
i. 7	c. 5	d. 3
g. 2	f. 9	a. 4

Yes!

What Does That Point Prove?

1. a. >
   b. <
   c. <
   d. <
   e. <
   f. >

2. a. 0.03, 0.033, 0.303, 0.33
   b. 0.007, 0.017, 0.02, 0.2
   c. 0.045, 0.4, 0.405, 0.45
   d. 0.04, 0.304, 0.32, 0.4
   e. 0.026, 0.06, 0.0602, 0.2
   f. 0.08, 0.082, 0.28, 0.8

Redundancy

1. a. 8
   b. 8
   c. 8
   d. 8
   e. 8
   f. 0.34
   g. 0.34
   h. 0.34
   i. 0.34
   j. 0.34
   k. 0.34

2. When each problem is put in standard division form, the problems are all the same.

3. Answers may vary.

Step Along

a. 8000, 1000, 1
b. 175, 6300, 10,000, 1
c. 9, 0.45, 1
d. 49, 350, 1000, 1
e. 1.2, 20, 1
f. 1.75, 0.21, 1
g. 72, 600, 1000, 1
h. 1.75, 21, 1

Four Straight

Estimates will vary.

Lesson 16: Changing Between Fractions and Decimals

Five-Minute Filler

Comedy: 0.5
History: 0.0833333
Drama: 0.25
Mystery: 0.16666667

Decimal Fractions

4. a. 0.75
   b. 0.666667
   c. 0.625
   d. 0.4
   e. 0.1666667
   f. 0.6363636
   g. 0.35
   h. 0.0833333
   i. 0.2857143
   j. 0.8
   k. 0.2777778
   l. 0.5555556

Completion Test

3. a. 0.1111111
   0.2222222
   0.3333333
   0.4444444
   0.5555556
   0.6666667
   b. 0.0909091
   0.1818182
   0.2727273
   0.3636364
   0.4545455
   0.5454546
   0.6363636
   0.7272727
   0.8181818
   0.9090909
   c. 0.1428571
   0.2857143
   0.4285714
   0.5714286
   0.7142857
   0.8571429
   d. 0.0833333
   0.1666667
   0.25
   0.3333333
   0.4166667
   0.5
   0.5833333
   0.6666667
   0.75
   0.8333333
4. Because the calculator rounds the last number up one if the following digits are > or = 5.

Order Up!

1. Answers will vary.
2. Answers will vary.
3. f, b, a, d, g, e, c
4. Answers will vary.
5. e, g, d, b, f, c, a

Converted

1. a. $\frac{3}{10}$
   b. $\frac{9}{20}$
   c. $\frac{689}{1000}$
   d. $1\frac{7}{20}$
   e. $12\frac{3}{25}$
   f. $\frac{19}{100}$
   g. $\frac{341}{1000}$
   h. $134\frac{1}{5}$
   i. $\frac{9}{100}$
   j. $3\frac{134}{200}$
   k. $\frac{16}{25}$
   l. $\frac{18}{25}$
   m. $2\frac{1}{4}$
   n. $7\frac{3}{4}$
   o. $12\frac{4}{5}$
2. a. 0.24
   b. 0.7143
   c. 0.5556
   d. 0.4167
   e. 0.7273
   f. 2.0794
   g. 0.5833
   h. 0.375
   i. 0.4286
   j. 2.2
   k. 1.2105
   l. 1.6667
   m. 0.2188
   n. 1.0690
   o. 0.2308

Converted (cont.)

3. e, c, d, b, a
4. e, d, c, a, b

Find the Fraction

5\. a. $\frac{1}{3}$ d. $\frac{41}{333}$

b. $\frac{5}{9}$ e. $\frac{2}{9}$

c. $\frac{65}{99}$ f. $\frac{3}{11}$

Lesson 17: Changing Percents to Fractions and Decimals

Five-Minute Filler

a. 50%
b. 75%
c. 100%
d. 125%
e. 150%

New Names

2\. a. 0.2, $\frac{20}{100}$, $\frac{1}{5}$ d. 0.375, $\frac{375}{1000}$, $\frac{3}{8}$

b. 0.9, $\frac{90}{100}$, $\frac{9}{10}$ e. 0.125, $\frac{125}{1000}$, $\frac{1}{8}$

c. 1.2, $1\frac{20}{100}$, $1\frac{1}{5}$ f. 2.4, $2\frac{40}{100}$, $2\frac{2}{5}$

100 Per

2\. a. 0.36 d. 0.79
b. 0.085 e. 13.41
c. 0.092 f. 1.51

Learn the Symbol

2\. a. 2.5% e. 406% i. 0.4%
b. 75% f. 20% j. 260%
c. 16.6% g. 194% k. 240%
d. 64% h. 50% l. 99.9%

3\. Move the decimal two places to the right and add the % symbol.

Percent Assigned

3\. a. 50% e. 22.2% i. 133.3%
b. 40% f. 137.5% j. 57.1%
c. 62.5% g. 24% k. 341.7%
d. 33.3% h. 275% l. 505%

Mixed Sequence

1. Calculator sequence:
   33%
   0.37
   55%
   $\frac{2}{3}$
   $\frac{3}{4}$
   0.80
2. Calculator sequence:
   $\frac{3}{8}$
   42%
   $\frac{1}{2}$
   65%
   0.75
   0.83

Lesson 18: Calculating with Percents

Sun Country

Scottsdale:	3.68%
Phoenix:	26.07%
Tempe:	4.46%
Mesa:	9.12%
Flagstaff:	1.99%
Tucson:	12.84%

One to Another

a. 60%
b. 88.9%
c. 40%
d. 250%
e. 40%
f. 25%
g. 30%
h. 80%
i. 153.8%
j. 2000%
k. 71.6%
l. 24.5%

Raises and Such

1. $518
2. $374.50
3. 5%
4. $156
5. 46 students
6. $268.80
7. $33.75
8. $3480.00
9. 0.9%
10. $3,337.50
11. 28%
12. $190

Commission er??

1. $2,625.00
2. $2,647.00
3. 17.8%
4. 7.3%
5. 10.3%
6. 254
7. 521
8. $1.95
9. $32.24, $528.24
10. 48.3%
11. $102.86
12. $4,800
13. 78.6%
14. 50

Lesson 19: Calculator Volumes and Circumferences

Five-Minute Filler

Volume = 8 cm^3
Surface area = 48 cm^2

Valid Relationships

Answers will vary.

Circle Up!

a. C= 12.57 ft
r = 2 ft

b. C = 94.25 cm
r = 15 cm

c. C = 18.85m
D = 6 m

d. D = 5 cm
r = 2.5 cm

e. D = 12 m
r = 6 m

f. C = 31.42 m
D = 10 m

g. D = 1.8 yd
r = .9 yd

h. D = 23 in.
r = 11.5 in.

i. C = 5.65 m
r = .9 m

j. 43.7 ft

k. 14.92 cm

l. 4.02 in.

The Yummy Yogurt Marketing Meeting

Container	Volume	Rank by Volume
A (cylinder)	19.79 $in.^3$	2
B (sphere)	17.97 $in.^3$	4
C (cone)	19.04 $in.^3$	3
D (cylinder + cone)	25.44 $in.^3$	1

Square "Roots"

Area	Radius	Diameter
	11 cm	22 cm
	9.04 cm	18.08 cm
221.67 cm^2		16.8 m
	11.07 cm	22.14 cm
	31.99 cm	63.98 cm
307,257.75 mm^2	312.74 mm	
1772.05 cm^2	23.75 cm	
0.35 m^2		$\frac{2}{3}$ m

Pizza Bargains

1. 40 cm pizza
2. two 30 cm pizzas
3. one 30 cm pizza
4. $8.89
5. $15.81

Varied Dimensions

a. 3287.52 m^3
b. 6575.03 m^3
c. 13,150.05 m^3
d. 821.88 m^3
e. 410.94 m^3
f. 205.47 m^3
g. 205,469.58 m^3

Lesson 20: Raising Numbers to a Power

Five-Minute Fillers

1. Answers may vary, but 729 is one.
2. No, Yes

Approximations

7a. 28.09
b. 11.18
c. 7.48
d. 32.33
e. 67.58
f. 186.52

Which Library Is This?

Bible

Who Was Pythagoras?

2. a. no
b. no
c. no
d. yes
e. no
f. no
g. no
h. yes
i. yes
j. yes
k. no

Power Match

3. a. 4
b. 5
c. 6
d. 2
e. 8
f. 3

4. a. 5
b. 3
c. 6
d. 2
e. 5
f. 6

Lesson 21: Reciprocals

Five-Minute Filler

smallest: 0.0000001
largest: 9,999,999

Upside Down

2. a. $\frac{1}{100}$
b. $\frac{1}{25}$
c. $\frac{1}{5}$
d. $\frac{1}{2}$
e. 1
f. 2
g. 5
h. 25
i. 100

Upside Down (cont.)

3. The product of a number and its reciprocal is 1.
4. You can't.

Over or Under?

a. $\frac{1}{12}$
b. 9
c. $\frac{2}{5}$
d. $\frac{1}{6}$
e. $\frac{7}{4}$
f. $\frac{8}{29}$
g. $\frac{1}{52}$
h. $\frac{100}{7}$
i. $\frac{3}{14}$
j. $\frac{1}{10,000}$
k. 64
l. $\frac{3}{31}$
m. $\frac{1}{28}$
n. $\frac{11}{9}$
o. $\frac{3}{71}$
p. $\frac{1}{195}$
q. $\frac{64}{47}$
r. $\frac{36}{3259}$

One Time!

a. 1
b. $\frac{1}{2}$
c. 2
d. 3
e. 4
f. $\frac{3}{2}$
g. $\frac{4}{5}$
h. $\frac{8}{9}$
i. 5
j. $\frac{4}{3}$
k. $\frac{7}{2}$
l. $\frac{5}{8}$
m. $\frac{8}{3}$

Increased Dividends

3. Same
6. They get larger quickly.
7. Get extremely large!

Lesson 22: Using Powers of 10

Five-Minute Filler

Over 3009 days

Notable Notations

2. a. 50,000
   b. 0.0045
   c. 360
   d. 0.073
   e. 900,000
   f. 0.000017
   g. 202.3
   h. 0.000901
   i. 50,400
   j. 0.0000602
   k. 101
   l. 0.04033

Notable Notations (cont.)

3. 750,000
   600,000
   1.419
   40
   0.0018
   0.0001
   159.903
   0.704

Lesson 23: Signed Numbers

Five-Minute Filler

-0.0000001

Learn the Signs

4. a. 4
   b. -14
   c. -4
   d. -38
   e. -8
   f. -251
   g. 339
   h. -200
5. a. subtract, larger
   b. add, the sign both have
6. a. -5
   b. 10
   c. 0
   d. -13
   e. 5
   f. -5

More Signs to Follow

2. a. 2
   b. -14
   c. 4
   d. -10
   e. -2
   f. 15
   g. -4
   h. 22
   i. 2
   j. -60
   k. 7
   l. 20
   m. 30
   n. 31
3. second number, addition
4. a. 3
   b. -6
   c. -12
   d. -1
   e. 27
   f. 40
   g. -8
   h. 5
   i. 2
   j. -6

Signs of the Times

2. a. -12
   b. 27
   c. -4
   d. 204
   e. -2925
   f. 1794
   g. 7830
   h. -5535
3. positive, negative

Signs of the Times (cont.)

4. a. -12   b. 14   c. 30   d. 70   e. 96   f. -120

5. They are the same.
   a. -4   b. 5   c. -18   d. 5   e. 14.93   f. 10.23

Signed On

1. -78.29° C
2. 5 sec
3. 16 ft

Lesson 24: Geometry Applications

Five-Minute Filler

Segments 10, 15, 21

Estimate the Area

(Estimates will vary.)

1. $A = 3406\ m^2$
2. $A = 1201.6\ cm^2$
3. $A = 180\ m^2$

Roaming Areas

1. a. 3888 m2   b. 3888 m2   c. 3999 m2   d. 3888 m2   e. 15,000 cm2   f. 15,000 cm2   g. 15,000 cm2
2. The sum of the bases does not change.

Long John Silver?

a. $b = 11.85$ cm
b. $h = 13$ cm
c. $A = 112.4\ m^2$
d. $h = 8.4$ cm

Prison or Prism?

2. $V = 1500\ cm^3$
3. a. $1500\ cm^3$ b. $1500\ cm^3$ c. $6000\ cm^3$ d. $375\ cm^3$ e. $93.75\ cm^3$ f. $40{,}500\ cm^3$

Lesson 25: Statistics

Average Magic

189.36	23.67	142.02	
71.01	118.35	165.69	It does!
94.68	213.03	47.34	

North to the Future?

1. 616,900
2. 5100

Mean, Median, and Mode

2. a. median = 42; mode = none; mean = 43
   b. median = $17,800; mode = $17,800; mean = $24,510
   c. median = 5.625; mean = 5.694
   d. median = 286; mean = 438.4
   e. mean price = 5.572

How Frequently Does It Occur?

Occurrence total: 154
Product total: 2010
Mean: 13.05

Lesson 26: Special Challenges

Five-Minute Filler

1. AB
   BA = 2! = 2

   ABC
   ACB
   BAC
   BCA
   CBA
   CAB = 3! = 6

ABCD	BACD	CABD	DABC
ABDC	BADC	CADB	DACB
ACBD	BCAD	CBAD	DBAC
ACDB	BCDA	CBDA	DBCA
ADCB	BDAC	CDBA	DCBA
ADBC	BDCA	CDAB	DCAB = 4! = 24

2. 5040
3. 362,880

Millionaires!

4. 2 yr 9 mo

Building Fences

458,366 square meters

Body Works!

1. 1440 min
2. 86,400 sec
3. 756,000 times
4. 1,209,600 times
5. 82.4 lbs
6. 20 cu ft
7. 29 bones

Math Explorer Overflow

a. 1,010,101
b. 133,333.33
c. 10,001
d. 1999.96

Only Two Keys Fit

a. 156
b. 390
c. 29
d. 1386
e. 762
f. 152

Did You Get That?

Answers will vary.

Check Your Answers!

3. 2
4. 9
5. 23
6. 10
7. yes
8. your first number

Calendar Magic

4. a. Answers will vary.
   b. Wednesday
   c. Wednesday
   d. Friday
   e. Friday

Temperature Conversion

3. 3.9
   176.7
   35
   -40